GENETIC BIOCHEMISTRY:
From Gene to Protein

ELLIS HORWOOD BOOKS IN GENE TECHNOLOGY

General Editor: Dr Alan Wiseman, Department of Biochemistry, University of Surrey, Guildford, Surrey

GENETIC BIOCHEMISTRY: From Gene to Protein
JACQUELINE ETIENNE-DECANT, Department of Biochemistry, Saint-Antoine Institute, Paris

GENETIC ENGINEERING
J. COLIN MURRELL and L. M. ROBERTS, Department of Biological Sciences, University of Warwick, Coventry

PRACTICAL ISOZYME GENETICS
N. PASTEUR, G. PASTEUR, F. BONHOMME, J. CATTALAN and J. BRITTON-DAVIDIAN, University of Montpellier 2, France

POLYTENE CHROMOSOMES IN GENETIC RESEARCH
V. SORSA, Department of Genetics, University of Helsinki, Finland

GENETIC BIOCHEMISTRY: From Gene to Protein

J. ETIENNE-DECANT
Professor of Biochemistry, Saint-Antoine Faculty of Medicine
University of Paris VI, France
in collaboration with
F. MILLOT
Tenon Hospital, Paris, France
Translation: JACQUIE WELCH
University of Southampton
Translation Editor: Dr J. L. FRANCIS
University Department of Haematology
Southampton General Hospital

ELLIS HORWOOD LIMITED
Publishers · Chichester

Halsted Press: a division of
JOHN WILEY & SONS
New York · Chichester · Brisbane · Toronto

This English edition first published in 1988 by
ELLIS HORWOOD LIMITED
Market Cross House, Cooper Street,
Chichester, West Sussex, PO19 1EB, England
The publisher's colophon is reproduced from James Gillison's drawing of the ancient Market Cross, Chichester.

Distributors:
Australia and New Zealand:
JACARANDA WILEY LIMITED
GPO Box 859, Brisbane, Queensland 4001, Australia
Canada:
JOHN WILEY & SONS CANADA LIMITED
22 Worcester Road, Rexdale, Ontario, Canada
Europe and Africa:
JOHN WILEY & SONS LIMITED
Baffins Lane, Chichester, West Sussex, England
North and South America and the rest of the world:
Halsted Press: a division of
JOHN WILEY & SONS
605 Third Avenue, New York, NY 10158, USA
South-East Asia
JOHN WILEY & SONS (SEA) PTE LIMITED
37 Jalan Pemimpin # 05–04
Block B, Union Industrial Building, Singapore 2057
Indian Subcontinent
WILEY EASTERN LIMITED
4835/24 Ansari Road
Daryaganj, New Delhi 110002, India

This English edition is translated from the original French edition *Biochimie génétique,* published in 1987 by Masson, Paris, © the copyright holders.

British Library Cataloguing in Publication Data
Etienne-Decant, J. (Jacqueline), *1933–*
Genetic biochemistry.
1. Genetic engineering. Biochemical aspects
I. Title II. Biochimie génétique, *English* III. Series
574.87'3224

Library of Congress Card No. 88–11202

ISBN 0–7458–0438–1 (Ellis Horwood Limited)
ISBN 0–470–21139–3 (Halsted Press)

Phototypeset in Times by Ellis Horwood Limited
Printed in Great Britain by Unwin Bros., Woking

Table of contents

Preface

Knowledge in this area of biochemistry, nucleic acids and their role in protein synthesis, has developed explosively in recent years. Following Crick and Watson, who opened the way to fruitful research, numerous other scientists have taken this direction: whole new areas such as oncogenes and genetic engineering have developed. Other topics such as the structure of DNA (and restriction enzymes, DNA sequences, Z-DNA, topo-isomerases, introns, mitochondrial DNA . . .) or viruses (hepatitis B virus, AIDS) are singularly substantial! It has not always been easy to choose which parts to cover. I should be grateful to all friendly readers of this first edition if they would not hestitate to let me have their criticisms. Further reading on all the topics covered is listed in the Selected Bibliography at the end of the book.

I very much appreciated the fact that Dr Françoise Millot was happy to 'transcribe' the figures from my tracings and to reread the manuscript. I thank Dr Francis Galibert who also did me the favour of rereading this text. I am very grateful to my husband for his constructive criticism.

Abbreviations

b.p.	Base pairs
aa	Amino acid
Ig	Immunoglobulin
ER	Endoplasmic reticulum
E.coli	*Escherichia coli*
A	Adenine
T	Thymine
C	Cytosine
G	Guanine
U	Uracil
cAMP	Cyclic AMP

(Note that in nucleotide chains, A, T, . . . signify 'A nucleotides, T nucleotides . . .')

1

Nucleic acids

I. GENERAL CHARACTERISTICS

Definition

'Nucleic acids' are substances which, as their name indicates, were initially isolated from the 'nucleus' of cells. In fact, as was discovered later, there are nucleic acids which exist not only in the nucleus, but also in the cytoplasm of cells. Thus although this term is no longer appropriate, it has nevertheless been retained.

There are two types of nucleic acids:

- DNA (deoxyribonucleic acid) localised essentially in the nuclei of cells (when the nucleus is individualised as in eukaryotes).
- RNAs (ribonucleic acids), found largely in the cytoplasm of cells.

Importance

The nucleic acids DNA and RNA play an essential role in protein synthesis in the following, rather schematic way:

- DNA contains 'the programme', i.e. the information needed for this synthesis. It will somehow serve in dictating the order in which particular amino acids form the chain to produce the final proteins. It also contains the information needed to regulate protein synthesis.
- RNAs allow this synthesis to be carried out.

Structure

Nucleic acids are very long molecules, formed by the repetition of sub-units called 'nucleotides').

II. NUCLEOTIDES

A. Elements constituting nucleotides

A nucleotide is itself made up of three elements:

Nucleotide=phosphoric acid+sugar+base

1. The base:
Two possible types of base are present in nucleotides:

- pyrimidine bases,
- purine bases.

Before studying these two types of bases, let us recall some basic formulae (Fig. 1).

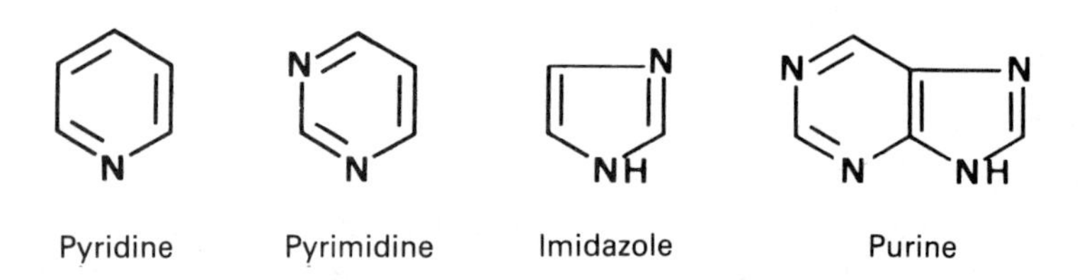

Fig. 1 — The C atoms and the H linked to C atoms are not represented (in these formulae and those following).

For simplicity and by convention, when representing these formulae, one does not write the carbon atoms (C) but merely the nitrogen atoms (N). (In this context, the nitrogen atoms should be written in the ring and not at the side).

Let us look first of all at the most simple bases:

(a) Pyrimidine bases
These all possess:

- a pyrimidine ring,
- various substituents which graft themselves onto this ring.

(Note that one part is common to all three formulae (Fig. 2): one OH between two N.)

In fact, for each of these three bases there is a balance between two tautomeric forms. For uracil for example (Fig. 3).

At physiological pH, the form C=O prevails over the C–OH form, whereas for example in cytosine, the C–NH_2 form prevails over the C=NH form. The three previous formulae of uracil, cytosine and thymine were only given in a didactic sense, since it is sometimes easier to arrive at the base formulae through the tautomeric C–OH forms. However, the three pyrimidine bases should then be written as shown in Fig. 4.

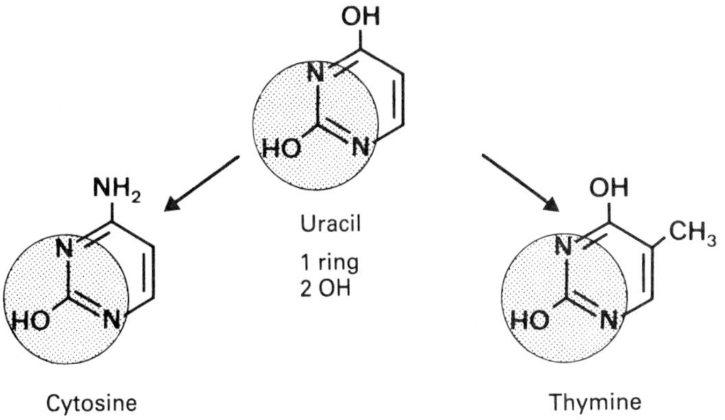

OH
NH$_2$
Uracil
1 ring
2 OH
OH
CH_3
Cytosine
Thymine

Fig. 2 — The pyrimidine bases (tautomeric forms C–OH).

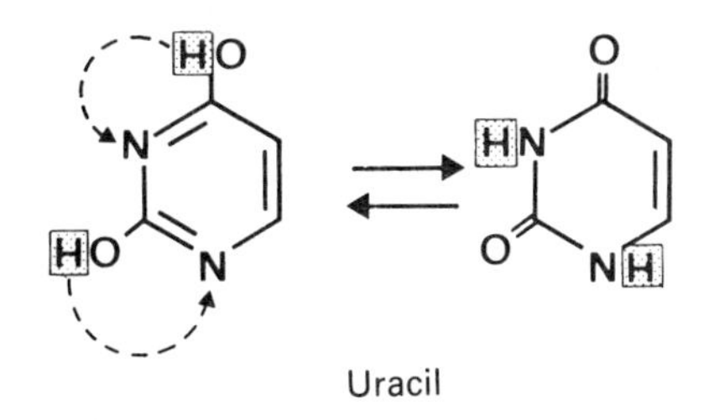

Fig. 3 — The tautomeric forms of uracil.

The numbering of pyrimidine bases has been specified simply to indicate the usage in a case where it would be necessary to refer to it.

(b) Purine bases
They all possess (Fig. 5):

- a purine ring,
- various substituents which graft themselves onto this ring.

We shall see that hypoxanthine does not form part of the five bases (A,T,C,G,U) incorporated in the nucleic acids during their synthesis. (It is, however, found in the tRNAs, where it will have been formed secondarily by deamination of the adenine of an AMP already incorporated into tRNA.)

The name 'guanine' comes from 'guano' (excrement from birds which is relatively rich in this base). The term 'purine' takes its origin from the word 'pus'. (In fact, in

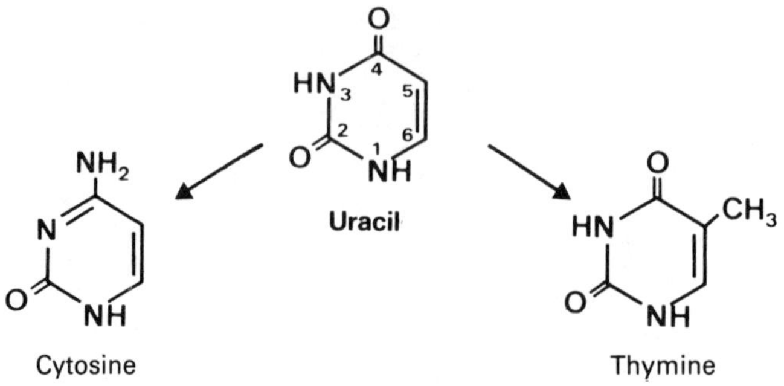

Fig. 4 — The pyrimidine bases (tautomeric forms C=O). At physiological pH, the form C=O prevails over the C–OH form.

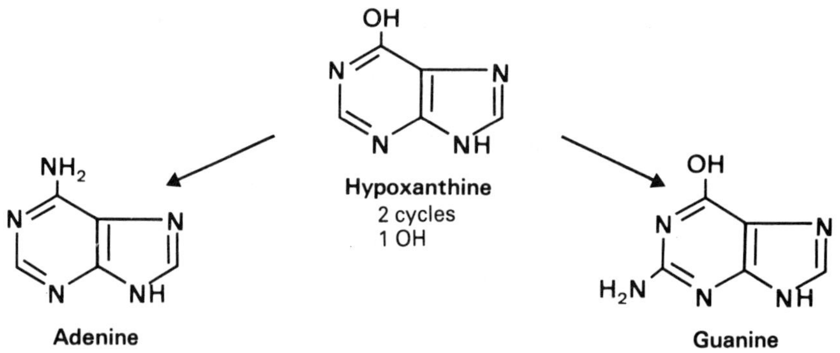

Fig. 5 — The purine bases (tautomeric forms C=OH).

1868, the Swiss chemist Miescher suspected cell nuclei of playing a role in heredity. He therefore studied cells with large nuclei. Thus, he was led to work on pus cells — that is to say white blood cells — which he obtained by washing purulent bandages (Fig. 6).)

In the case of purine bases it is also the tautomeric C=O form which, at physiological pH prevails over the C–OH form. The numbering of purine bases was also done for reference purposes.

2. *The sugar*

Two types of sugars are found in nucleic acids:

(a) Ribose (D-ribose)

Ribose is a monosaccharide at C_5 (Fig. 7).

This name comes from the initials of the institute where it was discovered: 'Rockefeller Institute of Biochemistry' in New York ('R.I.B.ose').

Ribose carbon atoms are numbered with 'prime' numbers to avoid confusion with the numbers of the bases.

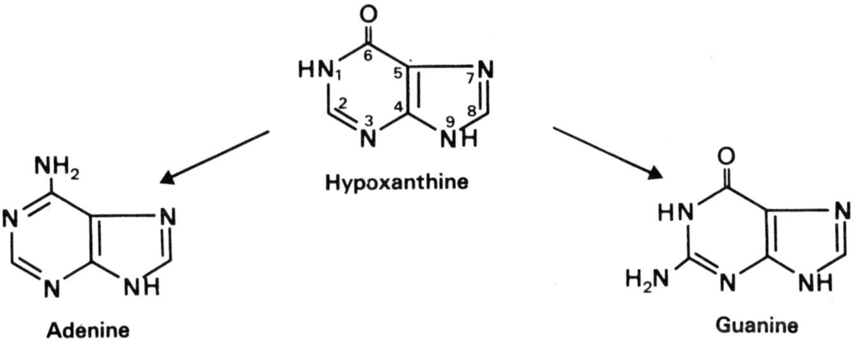

Fig. 6 — Purine bases (tautomeric forms C=O). At physiological pH the C=O form prevails over the C–OH form.

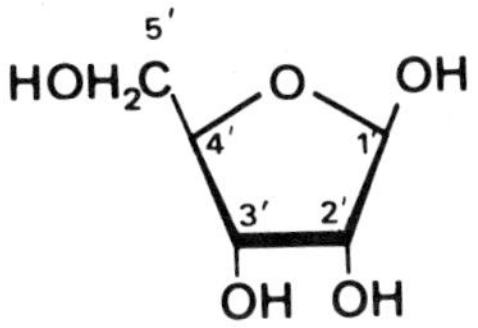

Fig. 7 — D-ribose. The C atoms at 1′2′3′4′, and also the Hs linked to these Cs have not been represented (either in this formula or the following ones).

(b) Deoxyribose (2′-deoxy–D-ribose)

Deoxyribose is a ribose in which an OH is missing at 2′ (this OH having been replaced by an H) (Fig. 8).

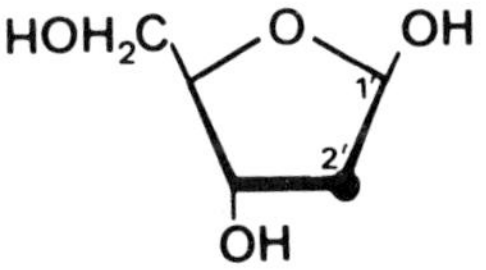

Fig. 8 — 2′-deoxy-D-ribose. (The absence of OH at 2′ has been marked by a solid circle in this figure).

3. *Phosphoric acid*

Phosphoric acid is a tri-acid. Two of the three acid functions will be esterified in DNAs and RNA (Fig. 9).

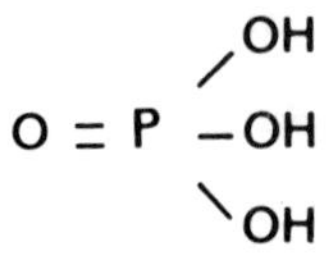

$$O = P \begin{matrix} \diagup OH \\ -OH \\ \diagdown OH \end{matrix}$$

Fig. 9 — Phosphoric acid.

B. Combination of the three elements constituting a nucleotide

1. *Sugar–base linkage*

The bond which unites the monosaccharide and the base is a β-osidic bond. This is formed by the elimination of a molecule of water between the semi-aldehydal OH of the monosaccharide and an H of the pyrimidine base (H at 1) or the purine base (H at 9) (Fig. 10).

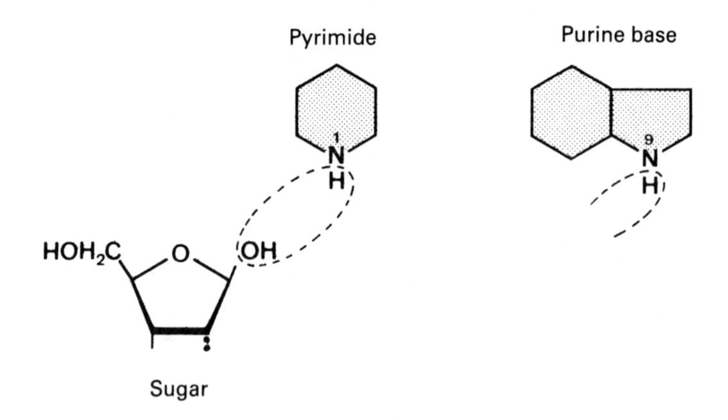

Fig. 10 — Sugar–base linkage.

This monosaccharide base grouping is called 'nucleo<u>s</u>ide' (not to be confused with nucleo<u>t</u>ide).

2. *H_3PO_4–sugar linkage*

The bond between the monosaccharide and phosphoric acid is an ester linkage. This bond is formed by the elimination of a molecule of water between:

- The OH of an acid: here it is the OH group of H_3PO_4

- The H of an alcohol: here it is the hydrogen of the alcohol function at 5′ of the monosaccharide (Fig. 11).

Fig. 11 — H_3PO_4–sugar linkage.

C. Names of the different nucleotides

1. Nucleotides containing a pyrimidine base

(a) Case 1: sugar is ribose

The nomenclature rule may be represented schematically as follows:

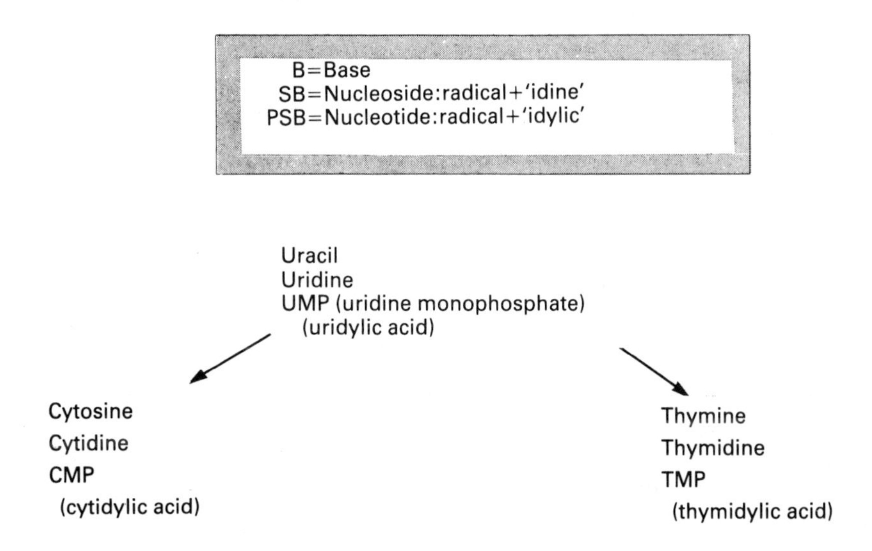

(b) Case 2: sugar is deoxyribose

For nucleotides where the sugar is a deoxyribose, the abbreviation for the nucelotide is prefixed by the letter 'd', e.g. dCMP, deoxycytidylic acid.

2. Nucleotides containing a purine base

(a) Case 1: sugar is ribose

The nomenclature rule can be represented in the same form as above:

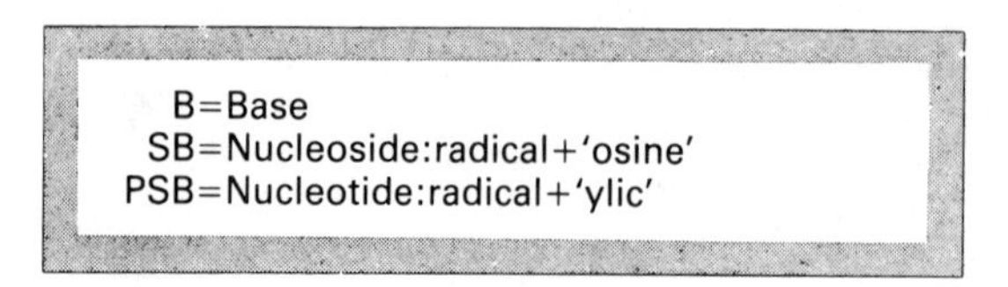

This nomenclature does not apply completely in every case: hypoxanthine and its derivatives, as can be seen below, represent an exception.

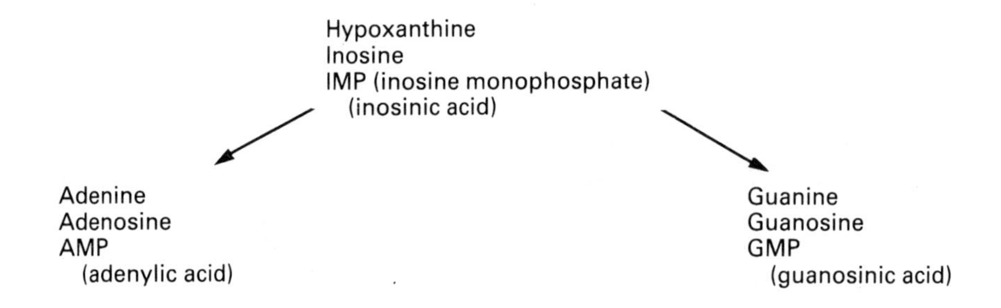

As we said before, IMP does not constitute part of the nucelotides seen in DNAs and RNA (with the exception of transfer RNA).

(b) Case 2: sugar is deoxyribose

In the same way as before a small 'd' is added when one wishes to specify that it is a deoxyribose nucleotide rather than a ribose nucleotide, e.g. dAMP, deoxyadenylic acid.

D. Combination of nucleotides in a nucleic acid

1. Bonds joining the nucleotides

In a nucleic acid the nucleotides are attached to each other by ester bonds. A molecule of water is eliminated between:

- The OH of an acid: here we have an OH group of H_3PO_4.
- The H of an alcohol: here we have the hydrogen of the alcohol function at 3′ of the monosaccharide.

Thus, phosphoric acid assumes two acid functions in bonds called 'phosphodiester' (one ester function serving to form the nucleotide, the second to combine two nucleotides.) The third acid function of H_3PO_4 remains free and thus confers acid properties on the nucleic acids DNA and RNA.

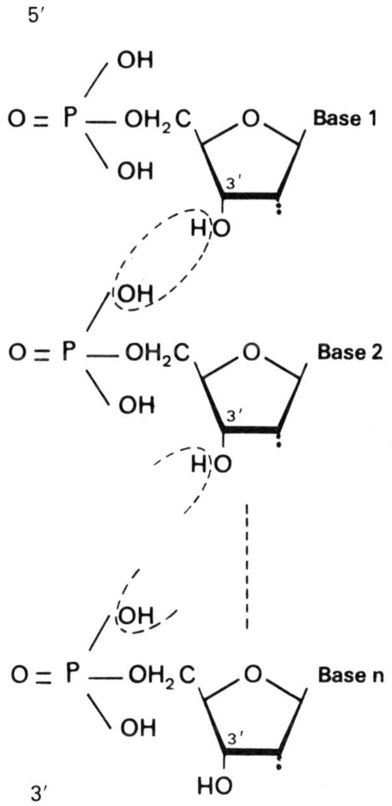

Fig. 12 — Chain of several nucleotides.

2. *The reading convention of a nucleic acid*

Just as, by convention, one reads proteins 'from the NH_2 end towards the COOH end', so there are reading rules for the nucleic acids.

Consider the two ends of a nucleic acid chain:

— One end contains the phosphate grouping with two free acid functions, and is called the '5′-P end'.

— The other end contains a free OH at 3′, on the monosaccharide, and is called the '3′-OH end'.

By convention, one always reads a nucleic acid chain from 5′-P towards 3′-OH.

In the case of a circular nucleic acid, one obviously no longer distinguishes a 5′-P end with two free acid functions and another end with a free OH grouping. However, the same reading direction is maintained, '5′ towards 3″'. For simplicity the method has now been adopted of having just the numbers 5′ and 3′ represented on each DNA sequence, whether it be a circular or a linear DNA.

III. DNA

A. Characteristics of DNA

DNA is formed from very many nucleotides linked together by ester functions.

1. Characteristics unique to DNA

Three characteristics are unique to DNA and differentiate it from RNA.

(a) The monosaccharide

As the initials 'DNA' indicate, the monosaccharide constituent of DNA is the deoxyribose (not the ribose as will be the case with the RNAs).

(b) The bases

The bases constituting the DNA nucleotides are:

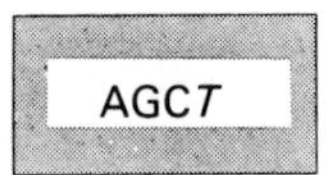

There are:

— two purine bases: adenine (A), guanine (G)
— two pyridimine bases: cytosine (C), thymine (T).

It should be noted that, in DNA, uracil (U) is never found (while in RNAs there is uracil in place of thymine). As we have already pointed out, DNA does not contain hypoxanthine.

(c) The two nucleotide chains

A molecule of DNA is usually formed from two chains (also called 'two strands') of nucleotides, whereas a molecule of RNA only has one. However, exceptions are noted for certain viruses.

2. Characteristics of the two DNA chains

These two chains have three essential properties; they are said to be:

- antiparallel,
- complementary,
- helicoid.

(a) Antiparallel

'Antiparallel' signifies that the two strands of nucleotides are parallel but in opposing directions. Taking the example of a linear DNA:

— For one strand, the direction '5′ to 3′' is found to be, for example, from top to bottom.
— for the second strand, the direction 5′ to 3′ will then be reversed, from bottom to top.

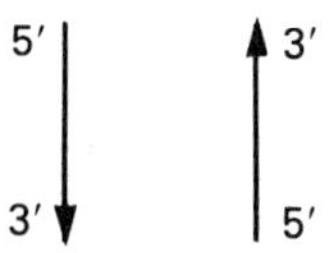

This can be written, for example as in Fig. 13. (For simplicity the base formulae are not represented).

(b) Complementary

• The rule of complementarity

The rule of complementarity is as follows:

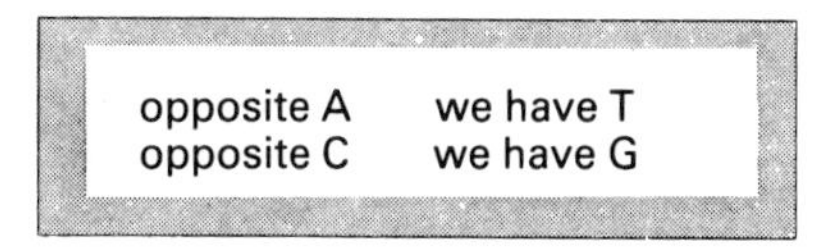

Thus if one knows the bases of strand I, those of strand II will be automatically known as well, thanks to the rule of complementarity. For example it may be:

5′A	T3′
C	G
G	C
A	T
3′C	G5′
strand I	strand II

This complementarity between the two chains is explained as follows:

For steric reasons (that is to say for reasons of space, or overcrowding), opposite to a purine, which consists of two rings, there is of necessity a pyrimidine, which has only one ring, and vice versa. Two purines (a total of four rings) would occupy too much space, and two pyrimidines (two rings in all) would be too elongated to form stable linkages, but with this arrangement each pair of bases has the same dimensions and this renders the regular structure of the double helix possible (Fig. 14).

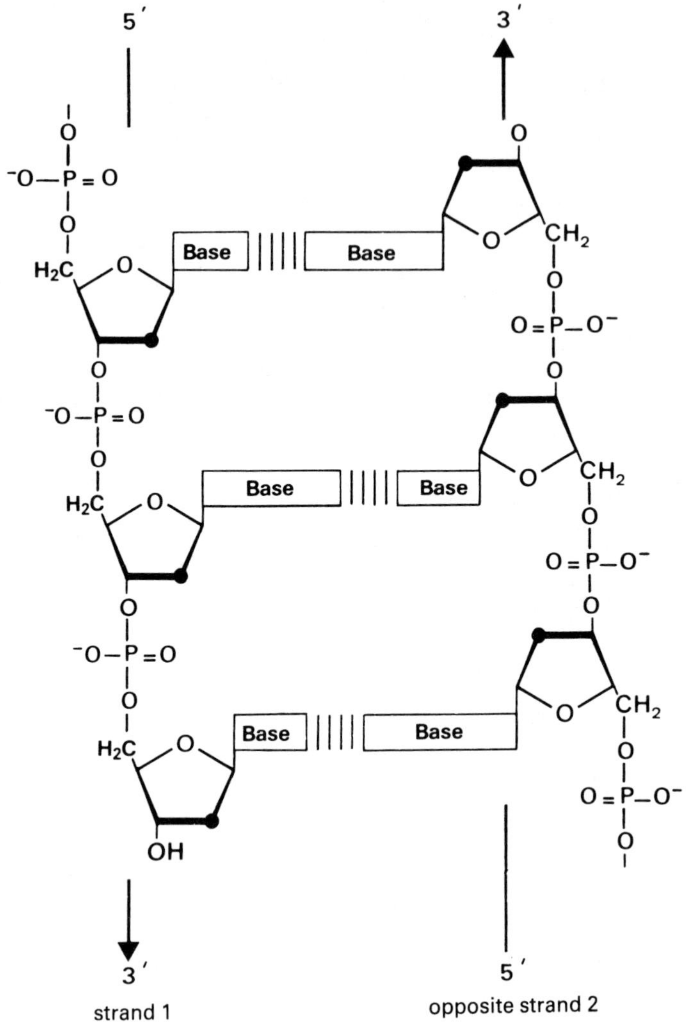

Fig. 13 — The two antiparallel nucleotide chains of a DNA.

This could lead to the supposition that opposite A (purine base) one could equally well find T as C (pyrimidine base). This is not so as we will now show.

Complementarity also results from hydrogen bonds. Face-to-face complementary bases are linked to each other by hydrogen bonds. Opposite an NH_2 grouping at C_6 of a purine base there is a C=O grouping of a pyrimidine base, and conversely, opposite a NH_2 grouping of a pyrimidine base there is a C=O grouping of a purine base.

Thus,

opposite A (6-aminopurine) there is T (4-cetopyrimidine)
opposite C (4-aminopyrimidine) there is G (6-cetopurine).

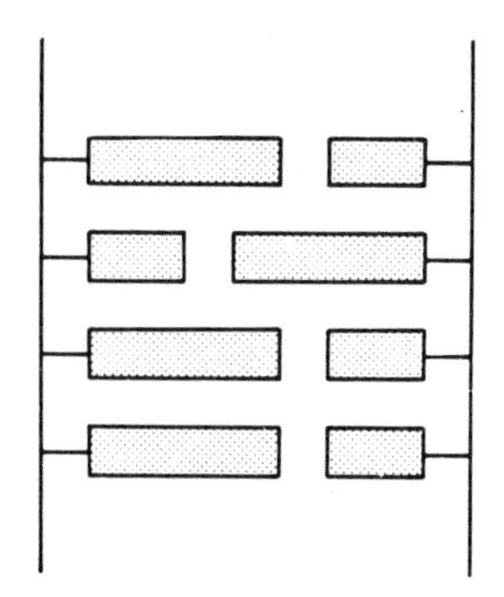

Fig. 14 — Each pair of bases of a DNA has the same dimension.

Moreover, for the C–G pair:

opposite grouping NH_2 at C_2 of G is the C=O grouping at C_2 of C.

Number of hydrogen bonds:
For pair A⁝⁝⁝T, it is two:

1 H between 1 N (covalent bond) and 1 O (hydrogen bond)
1 H between 1 N (hydrogen bond) and 1 N (covalent bond)

(The formula for A has been turned round here by comparison with the usual representation (Fig. 15)).

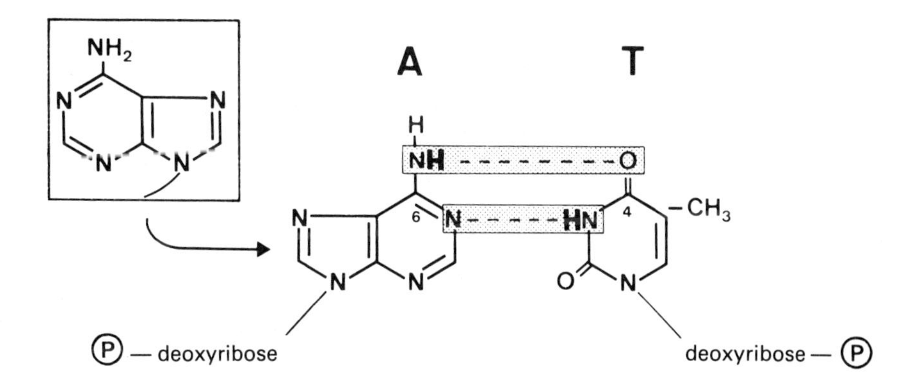

Fig. 15 — The two hydrogen bonds of the adenine–thymine pair.

For the pair C⁝⁝⁝G, it is three:

1 H between 1 O (hydrogen bond) and 1 N (covalent bond)
1 H between 1 N (covalent bond) and 1 N (hydrogen bond)

1 H between 1 N (covalent bond) and 1 O (hydrogen bond)

(The formula for G has also been turned round (Fig. 16)).

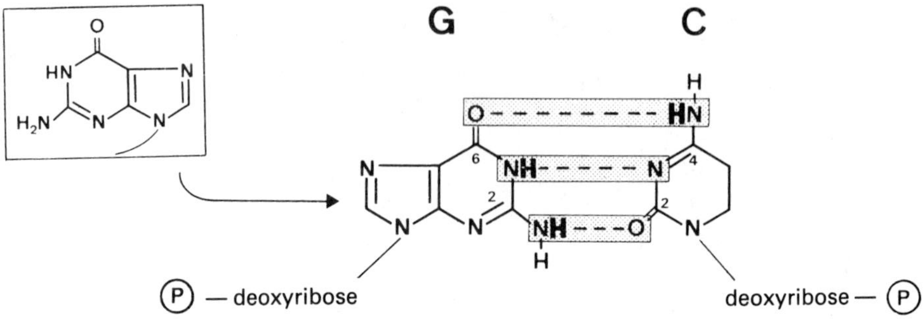

Fig. 16 — The three hydrogen bonds of the cytosine–guanine pair.

The presentation chosen here is very schematic. Its aim is to show as simply as possible which atoms are involved in these different hydrogen bonds. Quite obviously much more precise representations exist taking into account the true orientation of the bases in a double strand DNA molecule, and also the distances between the atoms. It should also not be forgotten that these formulae should be seen as occupying space.

In fact, two configurations of the C–N bond connecting each base to the sugar are possible, 'syn' or 'anti'. Put another way, there are two positions which a base can occupy with respect to a deoxyribose molecule. In DNA (A-DNA or B-DNA) all the purine and pyrimidine bases have the same conformation (anti). In another form of DNA (Z-DNA) we shall see that cytosine comes under the anti form, guanine under the syn form (there has been rotation of the base around the glycoside bond) and that the anti and syn conformations alternate (Fig. 17).

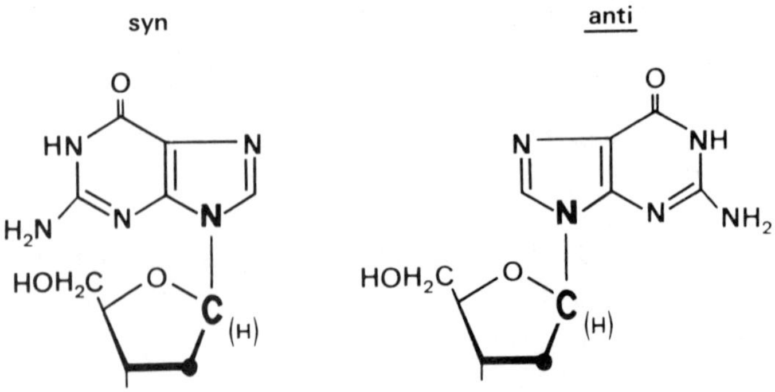

Fig. 17 — The two conformations, syn and anti of deoxyguanosine.

- Schematic representation

To summarise, one can represent a DNA molecule very schematically (without taking account of the spatial form) as a kind of ladder where (Fig. 18):

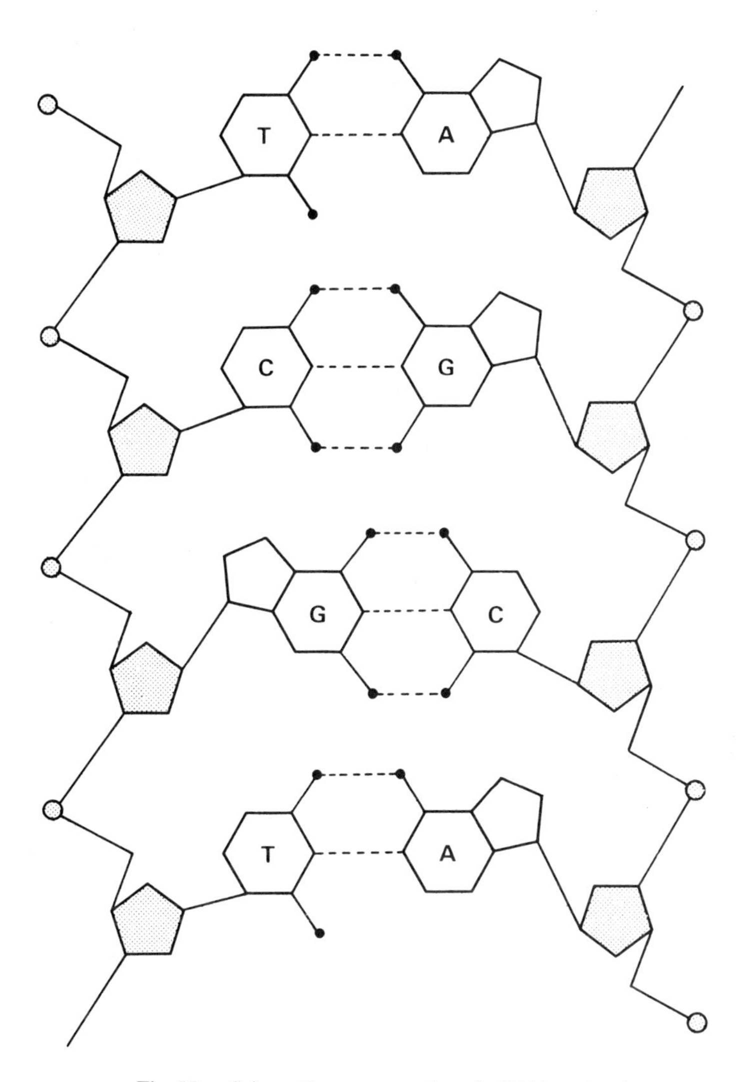

Fig. 18 — Schematic representation of a DNA molecule.

— the uprights are formed from deoxyribose and phosphoric acid,
— the bars are constituted by the complementary bases situated face-to-face.

- The sequence in DNA bases

One must understand that deoxyribose and H_3PO_4 are the same throughout DNA. This does not apply to the bases. As we have seen, there are four different bases. The order in which they occur, 'the base sequence', is characteristic of each DNA molecule.

By convention, to simplify, a nucleic acid chain is represented by writing down only the bases. They may be written vertically or horizontally: for example:

5' C	G 3'			
T	A			
A	T		5' C T A G C G G A 3'	
G	C	or	3' G A T C G C C T 5'	
C	G			
G	C			
G	C			
3' A	T 5'			

However, it should not be forgotten that in fact we are talking here of nucleotides and that these letters really signify: C. 'cytosine nucleotide', T. 'thymine nucleotide', A. 'adenine nucleotide', G. 'guanine nucleotide'.

- Comment on the denaturation of DNAs

If DNA is heated, a rupture of the hydrogen linkages between the bases is produced at a certain temperature, which is called, incorrectly, the 'fusion temperature'. The double helix loses strength and two strands separate: DNA is said to be denatured. the more C:::G bases there are in DNA the higher will be the fusion temperature. (This is easily understood, given that the C:::G bases are joined by three hydrogen bonds, whereas the A:::T bases are only joined by two hydrogen bonds).

The 'denaturation' of DNA (the separation of the two strands) is accompanied by important modifications of the physical properties. For example:

— viscosity decreases
— ultra-violet light absorption increases.

The denaturation is reversible: in certain conditions, when the temperature is lowered, the two strands of DNA can rehybridise (renaturation).

(c) Helicoid

The two chains of DNA adopt a helicoid configuration in space. They twine around an imaginary central axis, forming a right-handed double helix. (Further on we speak of another form of DNA (Z-DNA) which twists to the left).

Two forms of right-handed double helix DNA exist: A-DNA and B-DNA. These two types of DNA differ slightly in the diameter of the helix and in the orientation of the pairs of bases.

B-DNA is the most important biological form. The first physicochemical analyses were carried out on crystalline fibres of B-DNA. The base planes are perpendicular to the axis of the helix. These bases, with a hydrophobic skeleton, pile up one on top of another with a certain amount of displacement. Ten base pairs (b.p.) are found per spiral. The distance between two corresponding points on the thread of the helix (say 10 b.p.) is 3.4 nm. The diameter of the helix is 2 nm (Fig. 19).

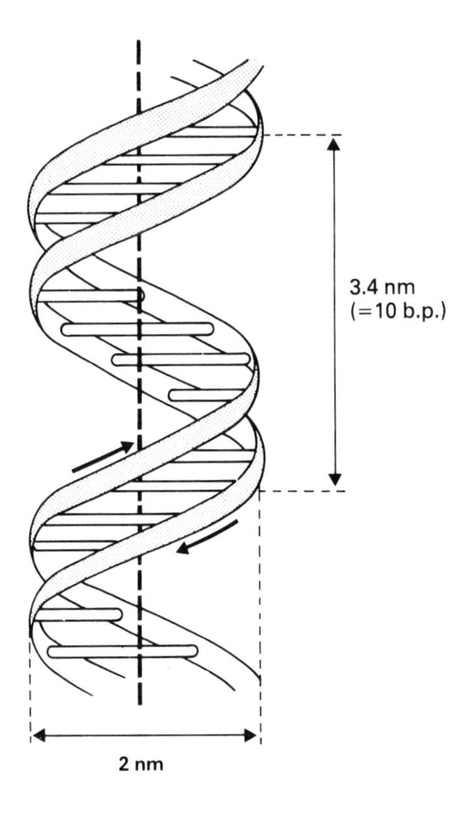

Fig. 19 — The double helix of DNA.

We shall see presently that, in living cells, DNA is found in solution and has a conformation near to that of B-DNA. All the same it possesses on average 10.5 b.p. per spiral rather than 10.

Examples of a double helix can be observed for example in the outside staircase of the Faculté des Sciences (Quai St. Bernard in Paris) or again the staircase of the Chateau de Chambord. However, at the Palais de La Découverte a giant molecular model exists which is much more exact.

This DNA structure was described by Crick and Watson in 1953. They received the Nobel prize for physiology and medicine in 1962 for this fundamental discovery which has since led to so many other discoveries.

B. The DNA of different living things

It is truly very remarkable that the DNA of all living things, animal, plant, bacterium or virus, possesses the same type of structure, two strands (with the exception of certain viruses) each constituted by a succession of several thousand nucleotides. What differs from one species to another will be:

- The number of DNAs in a cell (one molecule of DNA in *E. Coli,* several in higher beings).
- Its length: several thousand nucleotides (in a chromosome) or several thousand million (spread over a number of chromosomes).
- Its shape: linear or circular.
- Its location in the cell (separated or not from the cytoplasm by a nuclear membrane).
- The 'sequence of bases'. It is essentially this which will be, as we shall see, characteristic for each DNA. These sequences will play a primary biological role: different sequences of bases will give different messages.

1. Viruses
Viruses are the living things (if, however, they can be so considered, since they are incapable of reproducing or of synthesising themselves on their own) possessing the shortest nucleic acids. Viruses are distinguished in DNA and in RNA. The DNA (or DNA viruses) is formed from some *thousands* to several dozen thousand millions of nucleotides.

2. Prokaryotes
In the prokaryotes (except for *E. coli*) DNA is not located in a nucleus, but is found in the cytoplasm, where it constitutes a unique chromosome. This is circular in shape. Prokaryote DNA is longer — approximately one thousand times more so — than that of viruses. Thus, *E. coli* DNA consists of about 4 *million* pairs of nucleotides.

In prokaryotes, what are known as plasmids are sometimes found. These are tiny bits of DNA, also circular, which exist alongside the chromosome, and which are independent of the principal DNA. We shall look at these again in the chapter on genetic engineering.

3. Eukaryotes
In eukaryotes (entities having a nucleus in their cells) DNA is located in the nucleus.

Each chromosome contains a very long DNA molecule, the latter being folded and knotted. The total number of nucleotides in a human cell is very great, maybe approximately a thousand times greater even than in bacteria. One therefore finds about three *thousand million* pairs of nucleotides in the DNA molecules which constitute the 46 human chromosomes. This DNA is combined with proteins to form chromatin.

It should not be thought that a systematic relationship exists between the amount of DNA and the complexity of an organism. For example, certain Batrachia, a group of plants, have 30 times more DNA than man!

In eukaryotes one also finds DNA outside the nucleus. Thus mitochondria contain DNA. Knowledge about mitochondrial DNA has, moreover, increased greatly in recent years, as we shall see later.

C. New discoveries concerning the structure of DNA

1. *Topo-isomers and topo-isomerases*

(a) *Topo-isomers*

Definition

Two circular molecules of DNA with exactly the same sequence of bases may, however, differ in their number of twists, that is to say, the number of turns made by one strand around the other. These two DNAs, differing solely by the number of twists, are called 'topo-isomers'.

Different states of topo-isomers

— Slack state

In the condition called 'slack' the constraints or tension, on the double helix is minimal. It is the most stable configuration of the molecule. The first analyses carried out on crystalline fibres of DNA showed that this state is seen for 10 b.p. per turn. In fact, in living cells — and that is what particularly interests us here — DNA is in solution and it has been demonstrated that the state without constraint occurs with about 10.5 b.p. per helix (Klug, Nobel prize winner for chemistry 1982)

— overtwisted state

When DNA is found in an overtwisted state, two forms of overtwisting are theoretically possible:

positive overtwisting: the number of twists has increased.
negative overtwisting: the number of twists has decreased.

One must understand that a negative overtwisting corresponds in fact to an untwisting of the double helix (Fig. 20).

Whether it be negative or positive, overtwisting exercises a constraint on the molecule. This constraint sets up a spiralling, both with a positively and a negatively twisted DNA. (Those who have ever had problems untwisting their telephone cable will understand this! To obtain overtwisted DNA (positive or negative) requires an energy-producing input (which is logical since one passes from a state without tension to a state under tension (Fig. 21)).

In cells, DNA is found in two states (Fig. 22):

slack state (without constraint),
twisted state, resulting from a negative overtwist.

Topo-isomers can be separated by gel electrophoresis. In fact, the more a DNA ring is untwisted (negatively overtwisted), the more it is spiralled and thus compacted. Thus the more a ring is compacted the quicker it will migrate. The molecules of DNA can then be seen by irradiating the gel with ultra-violet rays, after colouring with ethidine bromide which produces an intense orange fluorescence (when it is combined with DNA in double helix.)

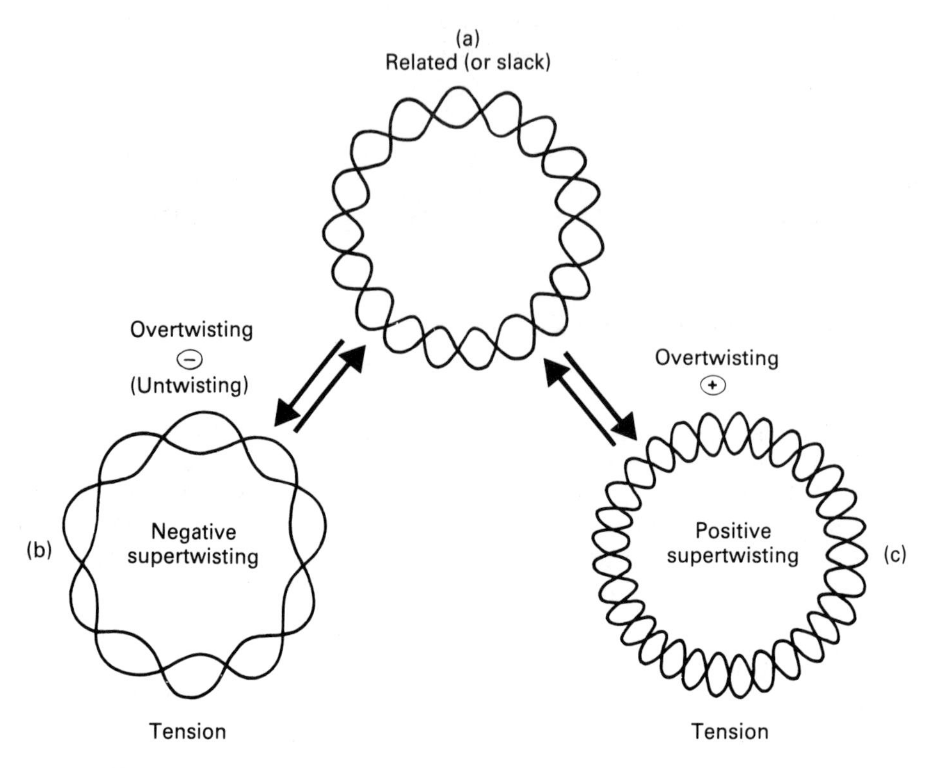

Fig. 20 — DNA topo-isomers: (a) slack state, (b) negatively overtwisted state, (c) positively overtwisted state. (Morange M. DNA's twisting enzyme. La Recherche, March 1980, p. 344, Fig. 1.)

It must be noted that these concepts of tension, of constraint, only apply theoretically to rings of DNA, and not to linear DNAs (where the ends are free to turn). Nevertheless a linear DNA can, like a circular DNA, be overtwisted if both the two ends are anchored to one point, as is the case in cells.

(b) Topo-isomerases

Definition

Topo-isomerases are enzymes which modify the number of interlacings. They are capable of introducing or eliminating supertwists in a DNA double helix. To modify the number of interlacings one or two strands of DNA must necessarily be cut.

The different topo-isomerases

Basically two kinds of topo-isomerases are described:

— Topo-isomerases I (e.g. relaxing enzymes)

These effect a transitory cutting and a re-fusion of a sole strand of bicatenary DNA. Relaxing enzymes are an example of this type of enzyme. They suppress the

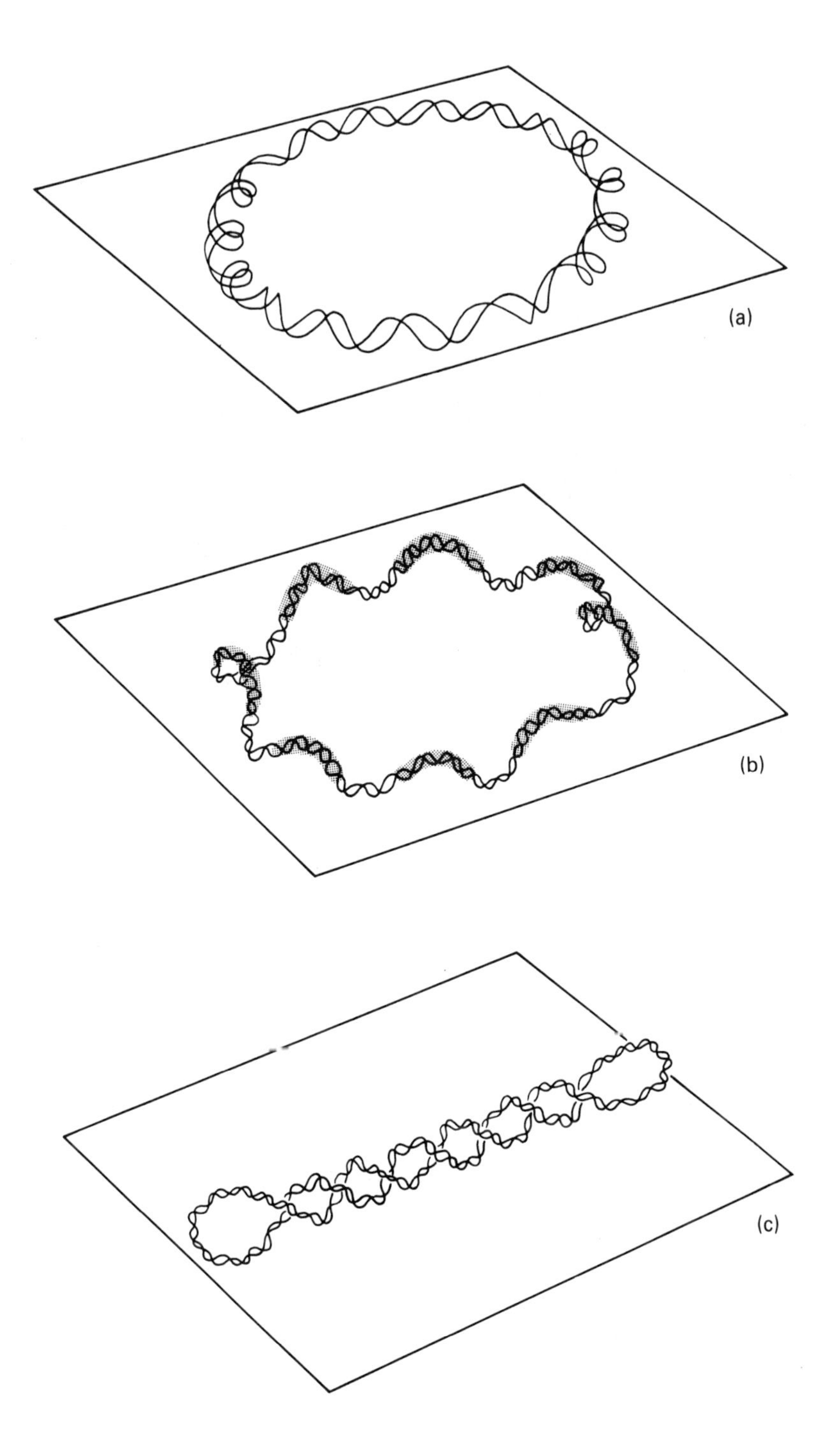

Fig. 21 — DNA topo-isomers: (a) slack state, (b) and (c) Negatively overtwisted forms. In (b) the helical skeleton cannot remain on a plane; and (c) shows formation of a superhelix: 'entwisted form'). (From Felsenfeld G. 'ADN'. *Science,* Dec 1985, p. 33, Fig. 3.)

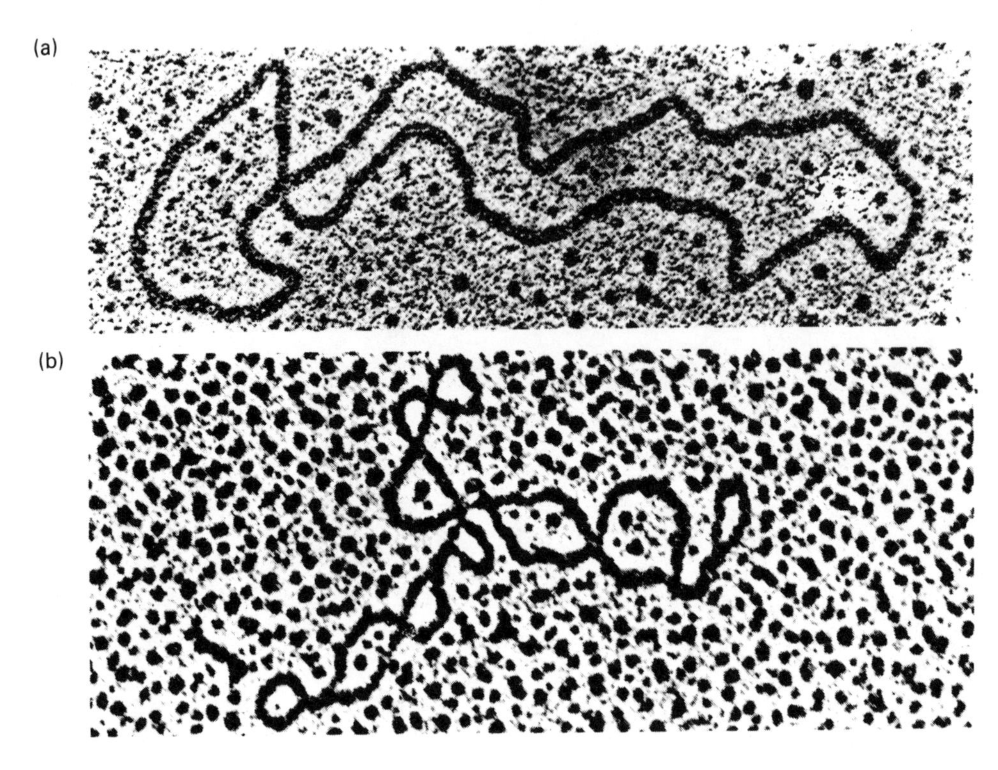

Fig. 22 — (a) Micrography showing a ring of slack DNA. (b) Micrography showing a ring of negatively overtwisted DNA (twisted). From Wang, J. Enzymes which modify the topology of ADN *Science*, Sept 1982, p. 59, Fig. 3.)

negative supertwists and bring back overtwisted DNA (under tension) to an initial conformation, that is to say, to the relaxing form (without tension). The action of these relaxing enzymes is as follows:

Firstly a DNA strand is cut with conservation of bonding energy by transitory formation of a DNA-topo-isomerase complex. In fact, a covalent bond is created between a phosphate of the DNA and a tyrosine from the active site of the topo-isomerase. The energy from the phosphodiester bond of the DNA is then temporarily conserved in this phosphotyrosine bond. The cut strand of DNA can then turn freely around the intact strand.

Secondly, the refusion of the DNA strand: the phosphodiester bond is reformed, with re-use of the previously conserved energy. DNA and topo-isomerases are thus regenerated, but the DNA has now acquired a configuration without constraint.

Studies on *E. coli* have shown that these topo-isomerases I operate essentially on negative supertwists. They are ineffective in relaxing positive supertwists.

— topo-isomerases II (e.g. gyrase)

These temporarily cut the two strands of DNA. An example is bacterial gyrase.

The gyrase is a remarkable enzyme which facilitates the introduction of negative supertwists in a double helix of DNA (then the unrolling of DNA). This reaction is coupled with the hydrolysis of a ATP molecule.

Biochemical importance of topo-isomers

- Viruses and prokaryotes

All the studies on topo-isomers and topo-isomerases have been carried out largely on bicatenary, circular, complete DNA molcules. These DNA rings are found in the majority of bacteria and viruses. The role of topo-isomerases seems very important:

— First of all DNA overtwists negatively, then spirals and is more compact than its circular relaxed equivalent. This spiralling enables a very long circular DNA molecule to be packaged in a small volume. Thus, *E. coli* DNA contains numerous loops. Each loop in fact comprises spiralled DNA.

— On the other hand, topo-isomerases facilitate replication or transcription (or recombination with other segments of DNA). In fact numerous proteins, indispensible from the start of these mechanisms, will only combine with DNA if it is negatively overtwisted, that is to say unrolled (for example, polymerase RNA after transcription).

The significance of gyrase capable of introducing negative supertwists can then be understood. The role of enzymes relaxing negative supertwists is more enigmatic. It is probable that they allow the degree of overtwisting to be adjusted. The overtwisting would then be obtained by a balance between the gyrase overentwisting negatively and the relaxing topo-isomerases.

- Eukaryotes

Topo-isomerases of type I and of type II have been detected in eukaryotes. They differ in several ways from relaxing enzymes and from the gyrase of bacteria. Their cellular function is not yet well understood.

As we shall see later, it is much more difficult to study DNA in higher organisms. In fact, the double helix of DNA twists around a collection of proteins, the histones, forming a superhelix. (However, it must be noted that mitochondrial DNA is not complicated by histones. It is only found in DNA that forms rings like that of bacteria.)

Medical application

Let us take as an example that of the mode of action of certain drugs such as urotrate or negram. These drugs are used in humans to treat colibacilli urinary infections. In part they would act in inhibiting the gyrase of the colibacillus. Negative overtwisting, normally produced by the gyrase and necessary for good unrolling in replication could not take place. The replication of DNA would therefore be impeded, and the colibacilli could not then multiply.

2. 'Left-handed' DNA or Z-DNA

(1) Presenting Z-DNA

Until recently the right-handed double helix of DNA had not been contested, but in

1979, Rich, in the United States, obtained a 'left-handed' DNA or Z-DNA. To obtain this result he had to utilise a double-stranded artificial oligonucleotide, of sequence (for one of the two strands): d(CGCGCG). This choice, where a pyrimidine and a purine base alternate may seem *a priori* arbitrary.

But such sequences, even if they are rare, exist *in vitro*. The need to use this type of artificial polymer does not constitute a major critique. Nevertheless the experimental conditions in which he had to operate were not physiological (so this medium had to be rich in salts). Since 1982 it has been possible to obtain Z-DNA in much more physiological conditions. It suffices, for example, to methylise these sequences of DNA (on the cytosine level) or to operate in the presence of certain polyamines such as spermine.

Finally, a very good experiment using specific antibodies directed against Z-DNA (which would only recognise Z-DNA and not B-DNA) enabled proof to be given, at the end of 1982, of the presence of Z-DNA in a substance of biological origin (insect chromosomes and also those of rodents). Z-DNA was no longer a laboratory curiosity — a year later it was also identified in human chromosomes.

(b) Characteristics of Z-DNA

This left-handed DNA is not a mirror image of right-handed DNA, it has a different conformation, for example:

- The sugar–phosphate skeleton is zigzag in appearance instead of forming a regular spiral as in B-DNA, (whence the name Z-DNA).
- Z-DNA forms a more slender helix with less twisted fringes than B-DNA. It contains a higher number of base-pairs per helix spiral. Thus one finds 12 b.p. per spiral (instead of 10.5 b.p. in B-DNA). The helix range is 4.46 nm (rather than 3.4 nm for B-DNA), and the diameter is 1.8 (instead of 2 nm for the B-DNA).
- As in B-DNA, the bases are situated inside the double helix. However, while they are perfectly inaccessible in B-DNA — anticonformation — they are exposed in Z-DNA — anticonformation for cystosine (or thymine) and syn for guanine (or adenine). It is the anti–syn alternation which produces the zigzag appearance of the Z-helix skeleton. On the same strand of Z-DNA the oxygen of the pentagonal cycle of the deoxyribose points alternately up and down, unlike in B-DNA.

Thus the imidazole nucleus of guanine (in particular N_7 and C_8) are clearly more accessible in Z-DNA.

A pyrimidine–purine alternation then is favourable to the Z-DNA conformation. The tendency for pyrimidine–purine dinucleotides to form Z-DNA sequences is, in descending order:

m^5CG CG TG=CA TA
(m^5C, cytosine methylated at C_5)

It would seem that prokaryote DNA forms more easily from Z-DNA than eukaryote DNA. This could be linked to the fact that the CG sequences are clearly

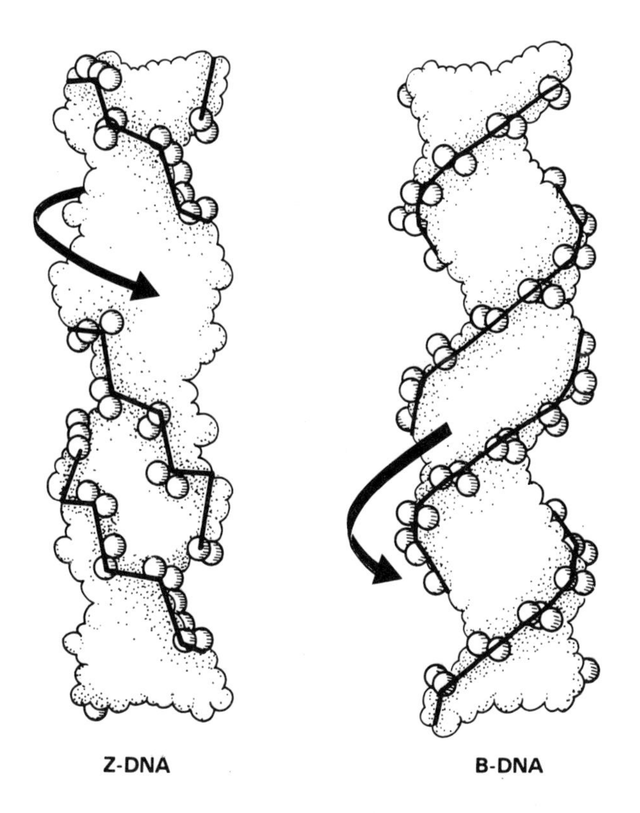

Fig. 23 — Z-DNA (left helix) and B-DNA (right helix) (Laigle A. When the double helix turns to the left. La Recherche, Oct 1982, p. 1188, Fig. 1.

more numerous in prokaryote DNA than in that of eukaryotes.

In physiological solutions, Z-DNA is less stable than B-DNA. This is primarily due to the electrostatic repulsions between the negatively charged phosphate groupings. These phosphates are closer in Z-DNA (more slender helix) than in B-DNA where the two strands are more elongated. This is why the initial observations were obtained with Z-DNA solutions rich in salts which reduced the phosphate–phosphate repulsion forces.

Thus DNA should no longer be considered as a static molecule, but rather as a dynamic, flexible, structure, where different conformations balance one another.

(c) Importance of Z-DNA

The exact function of Z-DNA is still not known. However, it is probable that it may have an important biological role. As we shall see later, the CG sequences seem to be implicated in the control of gene expression, at least in the higher eukaryotes. It would seem that there is an inverse relationship between the methylation of these sequences and gene expression. Thus, most frequently, these sequences would be methylated (on cytosine) in the inactive genes, and conversely, not methylated in the genes which are expressed. However, methylation of CG sequences strongly favours

Z-DNA formation. It remains to be seen if, *in vivo,* these CG sequences are capable of assuming, as soon as they are methylated, a Z-DNA conformation (double left-handed helix) at the centre of a double right-handed helix zone.

In addition, certain proteins are capable of combining specifically with Z-DNA, but not with B-DNA. It is probable that these 'Z-DNA binding proteins' have an important regulatory role.

Summary

We now discover that DNA contains not only 'coding' information which directs RNA synthesis (through the nucleotide sequence), but also 'conformational' information due to certain nucleotide sequences (for example, CG succession). This conformational information would then allow DNA segments to adopt differnt conformations (for example, Z-DNA) which could play an important role in certain biological regulations, in particular at the transcription level.

D. Study of DNA sequences

As we mentioned previously, two different DNA molecules have:

— the same molecules of phosphate and of sugar, but
— different bases. More precisely they differ in their base sequences, that is to say, the order in which all the bases of this DNA succeed each other. To establish the chemical formula of a DNA is to know the sequence in bases of this DNA.

Up until recent years, it was not known how to determine the base sequence of a DNA. It was only in 1976 that it was possible to effect the first sequence, from, obviously, one of the smallest molecules of DNA existing, that of a virus. This important step forward was obtained thanks to new techniques facilitating:

— the cutting of DNA in well-specified areas. It was possible to accomplish this, using enzymes called 'restriction endonucleases' or more simply 'restriction enzymes' (Arber, Smith and Nathans, Nobel prize for medicine and physiology 1978).
— The analysis of the DNA fragments obtained, by the enzymatic technique of Sanger or the chemical technique of Maxam and Gilbert. (Sanger and Gilbert both received the Nobel prize for chemistry in 1980).

1. Cutting of DNA by restriction enzymes

The restriction enzymes are really laboratory 'tools' used to cut DNA. The cutting is done at a particular site recognised by the enzyme. About one hundred restriction enzymes exist. Each one recognises a different nucleotide sequence which is specific to it.

(a) Which are the nucleotide sequences recognised by restriction enzymes?

The nucleotide sequences usually recognised by the restriction enzymes are called 'palindromes'. A 'palindrome' in literature is a word or phrase, which can be read equally well from left to right as from right to left, (for example, 'radar'). By analogy, in chemistry, DNA sequences with inverse symmetry are called palindromes, sequences such as the one framed below, where the succession of nucleotides is the same for strand I read from left to right (from 5′ to 3′) as for strand II read from right to left (likewise from 5′ to 3′).

```
5'... C C A|G A A T T C|C A T ... 3'     strand I
3'... G G T|C T T A A G|G T A ... 5'     strand II
```

Palindromes of every length exist (shorter or much longer), but the palindromes recognised by the restriction enzymes usually comprise sequences containing 4 b.p., or more often, 6 b.p.

(b) What is the origin of these restriction enzymes?
Restriction enzymes have been isolated in bacteria. In fact, as will be discussed later, the bacteria can be parasitised by DNA viruses. To defend themselves, the bacteria synthesise restriction enzymes (veritable 'pruning scissors') which will cut the DNA of these viruses.

Names which recall the origin of the bacteria from which they have been isolated are given to the restriction enzymes. For example, for indication purposes:

- Eco RI

Isolated from *E. coli* (strain RI). This enzyme recognises the palindrome:

```
5'...|G A A T T C|...3'
3'...|C T T A A G|...5'
```

- Hpa I

isolated from *Haemophilus parainfluenzae*. This enzyme recognises the palindrome:

```
5'...|G T T A A C|...3'
3'...|C A A T T G|...5'
```

- Hind III (or Hin dIII)

Isolated from *Haemophilus influenzae*, this enzyme recognises the palindrome:

```
5'...|A A G C T T|...3'
3'...|T T C G A A|...5'
```

(c) How does cutting with restriction enzymes take place?
Two types of cutting are observed:

- Cutting giving 'open ends'

Cutting of a DNA molecule can take place in the middle of a palindrome e.g. Fig. 24.

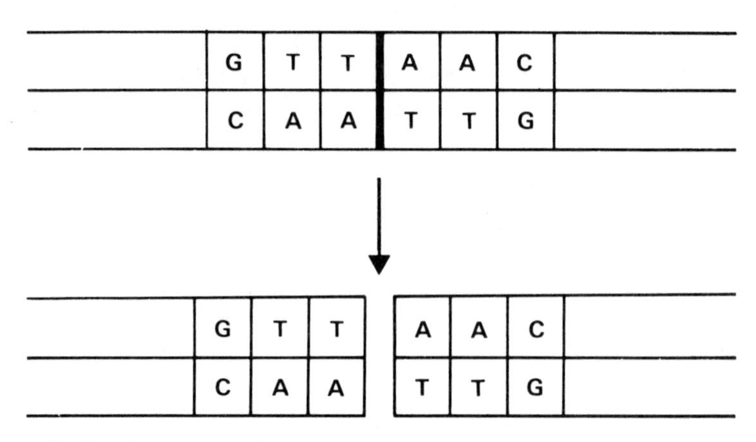

Fig. 24 — Cutting of a sequence of DNA by one restriction enzyme giving open ends (e.g. HpaI).

- Cutting giving 'adhesive ends' (or 'sticky tips')

Other types of enzyme act in cutting from one part and another of the centre of symmetry (e.g. Fig. 25).

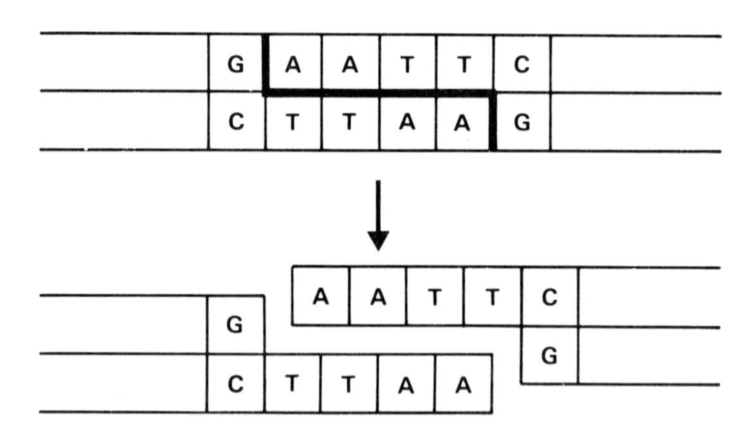

Fig. 25 — Cutting of a sequence of DNA by one restriction enzyme giving adhesive ends (e.g. Eco RI).

Later we shall see the significance of these two types of cutting in the chapter on the techniques of genetic engineering.

A restriction enzyme is therefore capable of making guiding marks the whole length of a DNA sequence representing certain characteristic sequences. Let us go back to the example of the enzyme (Eco RI) which recognises the sequence

```
   . . .|G A A T T C|. . .
   . . .|C T T A A G|. . .
```

If this sequence occurs five times in a molecule of DNA, the enzyme will cut this DNA in those five places. The DNA will then yield five fragments if it is a circular DNA (or six in the case of a linear DNA). Quite clearly, another restriction enzyme will recognise another sequence different from that recognised by Eco RI. According to the enzyme utilised, an identical DNA will be cut differently.

2. Determination of the base sequence from fragments obtained from DNA

After the cutting of a DNA molecule by restriction enzymes, frequencies are obtained which have, for example, a hundred nucleotides, for which the base sequence can then be determined.

The recent techniques proposed by Sanger, or by Maxam and Gilbert, have enabled extraordinary progress to be made. We shall only give here, by way of example, the principle of Maxam and Gilbert's technique:

(a) Marking for each of the two strands of the DNA molecule at ^{32}P (end 5′)

To achieve this marking, it suffices to sample the last phosphate at 5′ with the help of a phosphatase, then to reintroduce a radioactive phosphate (using ATP, the last phosphate of which is radioactive ([γ-^{32}P]ATP) and a kinase).

(b) Separation of the two DNA strands

On heating DNA, the hydrogen bonds between the two strands are destroyed. (As we saw previously, it is said that a denaturation of DNA is carried out). These two strands can then be separated by electrophoresis. Analysis is carried out on one strand of DNA.

(c) Chemical cleavage

The fragment, the sequence of which is being determined, will be split at the level of a well-defined base which will then produce many small fragments. For example, dimethyl sulphate is a reactive which methylates guanine N_7 and adenine at N_3. The glycoside bond of a methylated purine is unstable and is easily broken by heating, thus leaving the semi-aldehyde function of the monosaccharide free. However, in the sugar–phosphate skeleton, the place at which there is a monosaccharide not linked to a base represents a weak point. Thus, monosaccharide–phosphate splitting (from the skeleton) takes place easily here and there in this monosaccharide (at the level of ester linkages at 5′ and 3′). By playing on these in operating conditions, cleavage at G and at A can be promoted.

For cytosine and thymine, the reactant initially used is hydrazine.

As we have said, this technique can be applied to determine the sequence of a fragment of DNA possessing a hundred molecules. To simplify, we shall give an example with a fragment having only seven nucleotides.

5′ *GCTAGTC3′

(the marking of the 5′ end by ^{32}P has been represented by an asterisk).

The different cleavages are carried out simultaneously on four samples from an identical segment of DNA.

Cleavage at A. The small fragment obtained is:

*GCT (situated on the left of the split) — three nucleotides

The GTC fragment, at the right of the split, is obtained in the same way. However, this fragment is not radioactive. It cannot, therefore, be observed later.

Cleavage at G. Obtained here:

*GCTA — four nucleotides

The comment above also applies to the non-radioactive fragments CTAGTC and TC.

Cleavage at C. Obtained here:

*G — one nucleotide
*GCTAGT — six nucleotides

Again non-radioactive fragments are obtained.

Cleavage at T. This split gives the following fragments:

*GC — two nucleotides
*GCTAG — five nucleotides

Again non-radioactive fragments are obtained.

For each of these cleavages an objection immediately springs to mind. Take for example the cleavage at T. We have said that this cleavage produced *GC and *GCTAG. Why do we get *GCTAG? In fact it is anticipated that a split will occur systematically at the level of each of the Ts, and consequently also at the level of the T of *GCTAG. It is important to understand that these cleavages are effected in conditions where the reaction takes place gently and not abruptly. It is said to be limited, which means that the reaction is only allowed a quite brief time during which to occur. Thus, the cleavage at the level of a base (let us continue with the example of T), will take place randomly for different Ts and not systematically at the level of all the Ts. Since very many molecules exist per sample (several billion) all the possibilities for cleavage at T are bound to be represented.

(d) Migration on a sheet of gel

After cleavage, the various fragments obtained will be separated according to size. This separation is done by electrophoresis on a sheet of gel: the lower the molecular weight of a fragment the further it will migrate.

The non-radioactive fragments also migrate according to their size, but they will not be marked in the following phase.

(e) Revelation of radioactive fragments after migration

After migration, all the radioactive fragments will be shown up by the application to the gel, of a film sensitive to the beta particles emitted by ^{32}P. Each radioactive area of gel will produce a spot on the film. The result obtained may be represented by the diagram in Fig. 26.

5′ 3′
*GCTAGTC

Cleavage at:

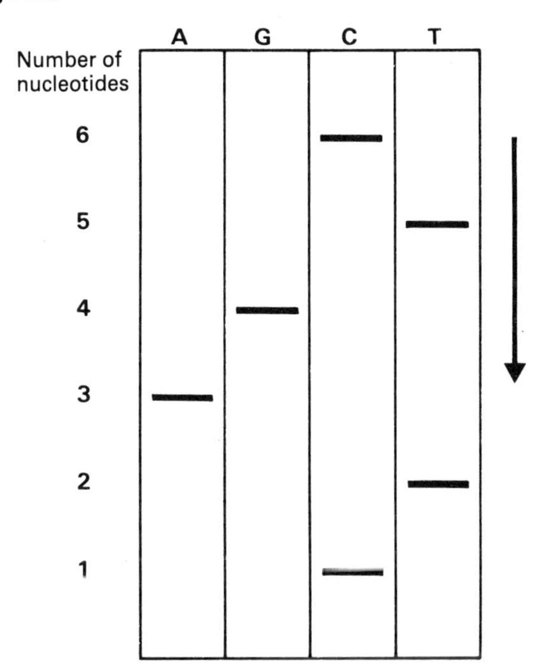

Fig. 26 — Diagram of the chemical cleavage of a segment of DNA with one strand: cleavages at A,G,C,T.

The radioactive fragments marked are for:

- Track A: the fragment with three nucleotides produced from cleavage at A.
- Track G: the fragment with four nucleotides produced from cleavage at G (its migration is slower than the previous three nucleotide fragments).
- Track C: the fragments with one and six nucleotides obtained from cleavage at C.
- Track T: the fragments with two and five nucleotides resulting from cleavage at T.

(f) Interpretation of a migration diagram

Let us examine the visible spots, track by track, with the aim of reconstituting the sequence in bases of the fragment with the seven nucleotides initially analysed.

- Track of the cleavage at A. The spot obtained is in position 3, which signifies that a fragment with three nucleotides occurs at the left of an A base. We can thus write A in the fourth position:

  ...A

- Track of the cleavage at G. The spot obtained is in position 4, which signifies that a fragment with four nucleotides occurs to the left of a G base. We can then write G in fifth position:

  ...AG

- Track of cleavage at C. Two spots are visible, one at 1, the other at 6. This indicates that a fragment with one nucleotide and a fragment with six nucleotides occur to the left of C, which allows us to write C in the second and seventh positions:

  .C.AG.C

- Finally, track of cleavage at T. Two spots at 2 and 5 are seen, which indicates that the T base occurs in the third and sixth positions, which gives finally:

  .CTAGTC

In practice, very simple, it suffices to read the spots from bottom to top, simultaneously on the four tracks, without it being necessary to apply the above reasoning. One thus obtains the base sequence (of 5′ towards 3′) very quickly.

By this technique, we have thus been able to determine the sequence of the fragment analysed, except for the first base. (in fact, it is the same two first bases of side 5′ which — for technical reasons — escape this investigation.) It suffices for this to determine in a similar way the sequence of the complementary strand. This makes it possible on the one hand to confirm the sequence of the first strand, and on the other to obtain the nature of the two first bases at 5′ (if one knows the two bases of the 3′ end of strand II, one automatically deduces from them, by applying the rules of complementarity, the two bases opposite them at the 5′ end of strand I).

```
5'xx__________________..3'
                            strand I
3'..__________________xx5'  strand II
```

Where xx represents bases not determined directly but by deduction in applying the rule of complementarity, and .. represents bases determined directly.

In a second experiment, a second example of DNA was cut by other restriction enzymes at other points. Thus a new lot of fragments was obtained whose sequence was determined. After a third and, if necessary, a fourth trial, one will finally elicit, thanks to overlapping zones, the total sequence of the DNA analysed initially. A real puzzle....

Let us complicate it a little more. The technique presented above has been simplified for reasons of clarity of exposition. In fact, reactants giving a cleavage solely at the level of a base are not generally utilised. To allow for re-cuttings it is preferable to use a reactant which produces a cleavage at the level of two bases, but where one of the two cleavages is greatly preferred.

For example,

1. Cleavage at A G.

The cleavage is not only at A, but at A and G. Nevertheless the cleavage at A will be much more rapid than that at G, to the extent that the spot at A will be more intense than that at G.

2. Cleavage at G A.

A cleavage solely at G is not utilised but one at both G and A. The cleavage at G being more rapid than that at A, the spot at G will be more intense than that at A.

3. Cleavage at C.

A reactant is used which produces a cleavage only at C.

4. Cleavage at C+T.

Lastly, a cleavage at T only is not used, but at C and T together. The interpretation of this track is a little more delicate: one cleavage at C will give a spot on track C (but equally on track T); one cleavage at T will give a spot only on track T.

Having shown the diagrams (Fig. 27), let us now see the photographs of the films actually obtained (in fact they are partially enlarged) (see Fig. 28).

This figure is taken from the article by Gilbert in *Proc. Natl. Acad. Sci.* (1977, **74** 563), where he published his famous technique. This served him in determining the sequence of a small segment of DNA from *E. coli* (taken from the *lac* operon).

Other types of cleavage could equally well be utilised: for example, at G, at G+A, at C+T, at C, at A C. Fig. 29 is taken from an article which appeared in *Nature* (1979, **281** 647) where the authors (Galibert *et al.*) had determined the sequence of a fragment of DNA from the hepatitis B virus.

Thus numerous nucleotide sequences are now known, using the method of Maxam and Gilbert, or that of Sanger, where the principle is a little different (enzymatic method or dideoxynucleotide method).

The first complete DNA sequence published was, in 1977, that of a small virus ϕX174 (pronounced 'phi X174'). This virus which parasites bacteria is formed from a single strand of circular DNA containing 5375 nucleotides (in fact some time later a

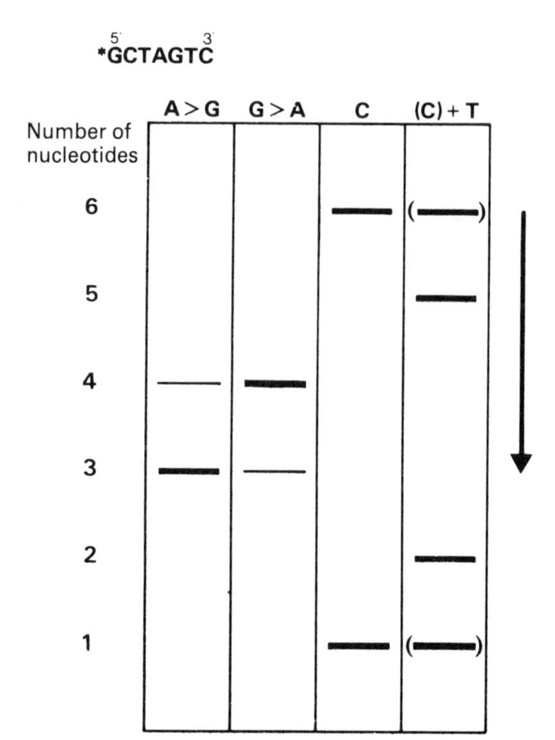

Fig. 27 — Principle of Maxam and Gilbert's method: chemical cleavage at A>G, G>A,C,C+T.

correction was added taking to 5386 the number of nucleotides of ϕX174).

The DNA sequence of *E. coli* is not yet known, but, as it comprises about 4 million b.p., a whole book of about 800 pages written in extremely fine print would be required to print it. As for the DNA of a human cell, if the nucleotides from 23 pairs of chromosomes were totalled up it would not be a book but a work of 1000 volumes which would be needed...

E. Modification enzymes

Modification enzymes are enzymes possessed by bacteria to protect against autodestruction. As we have already seen, bacteria use their restriction enzymes to cut the DNA of a virus attempting to parasitise them. Given that bacteria also contain DNA, the restriction enzyme which they synthesise to combat a virus must not also cut their own DNA. Every bacterium (not being suicidal!) has therefore to 'modify' its palindromes so as not to be recognised by its own restriction enzymes. The modification effected is a simple methylation, thanks to enzymes called 'modification methylases', or 'methylation enzymes'. This methylation is produced on a base situated in the palindrome. Let us take the example of the restriction enzyme Hind III which, as we have seen, recognises the palindrome:

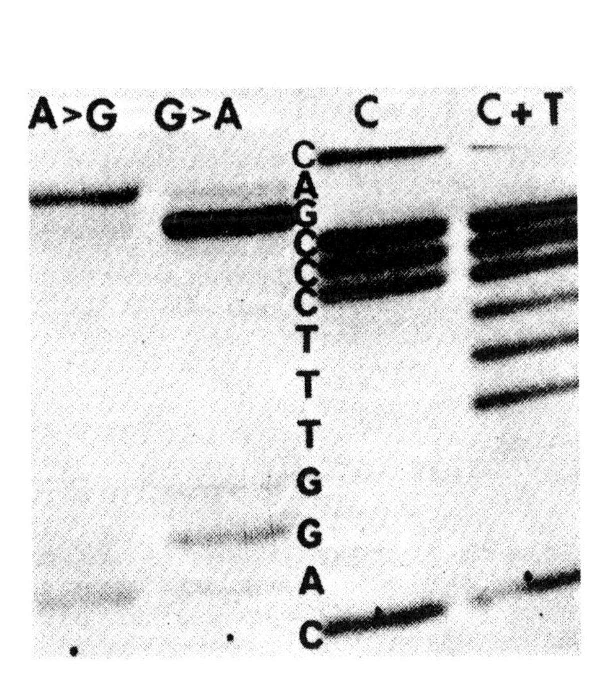

Fig. 28 — Autoradiogram of a gel after sequencing with Maxam and Gilbert's method (only a small portion of this gel is shown).

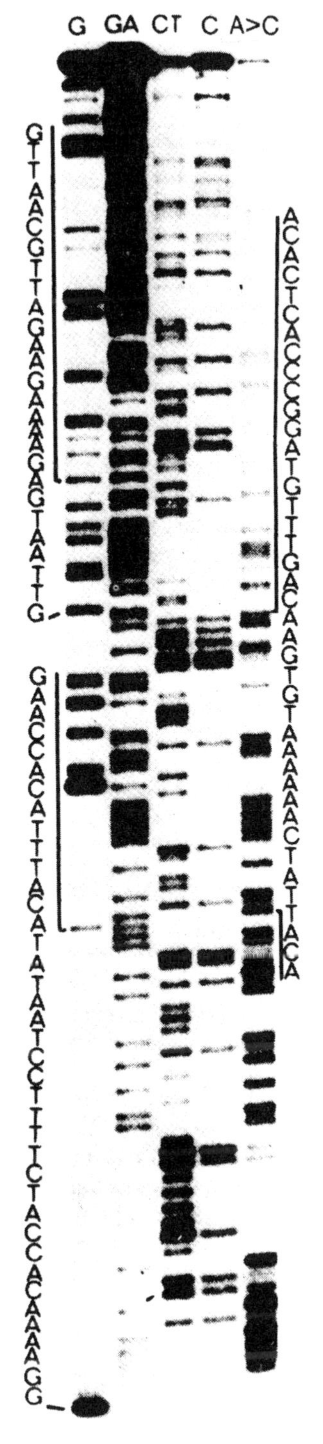

Fig. 29 — Determination of the sequence of hepatitis B virus using the Maxam and Gilbert method (nucleotides 552–658). (Galibert F. *et al.* 'Nucleotide sequence of the heptatis B virus genome' (Subtype ayw) cloned in *E. coli*. Nature Oct 1979, p. 647, Fig. 1.1.)

```
     *
...|A A G C T T|...
...|T T C G A A|...
              *
```

After methylation of adenine (represented here by a solid circle), this palindrome is no longer recognised by restriction enzyme Hind III and so can no longer be cut.

Comments

- In the laboratory, restriction enzymes are used which specifically recognise a palindrome. However, it should not be deduced from this that all known restriction enzymes act systematically at palindrome level. It is now known that certain strains of bacteria possess other types of restriction enzymes.

  These enzymes recognise particular sequences which, however, do not possess any symmetry, e.g. the sequence:

  TGA........TGCT

  (where each of the eight points represents any of the four bases).

  Moreover, these enzymes cut DNA, not at the level of the sequence which they have recognised, but at a place which may be located at about 1000 b.p. from this site!
- There are even restriction enzymes capable of cutting viral DNA and methylating bacterial DNA at the same time. (These are enzymes possessing several sub-units.)

IV. THE RNAs

A. Characteristics of the RNAs

The RNAs (ribonucleic acids) are characterised essentially by:

- the monosaccharide: as the name (RNA) indicates, the monosaccharide is ribose, unlike DNAs where the sugar is deoxyribose.
- The bases: the bases found in RNAs are A, C, G, and U in place of T.
- A single chain of nucleotides (and not two chains as in DNA). This chain is moreover much shorter than the DNA chains.

B. The rules of pairing

The rules of pairing between two strands of RNA will be the same as between two strands of DNA as far as C and G are concerned; however U replaces T in pairing with A:

A : : : U
C ⁝ ⁝ ⁝ G

Pairing between complementary bases can then be observed either between two different molecules of RNA (since one molecule of RNA only has one strand), or on an identical molecule of RNA, in an area bent into a horseshoe shape (an area where a 'self-complementary' exists).

C. The different RNAs

Cells contain three main types of RNA:

— ribosomic (rRNA),
— transfer (tRNA),
— messenger (mRNA).

Later we shall have occasion to discuss a fourth type of RNA, the snRNAs.

It is the rRNAs which are by far the most abundant RNAs of the cell (about 82%), whereas the tRNAs only represent about 16%, the mRNAs about 2% and the snRNAs less than 1% of the total cellular RNA.

1. Ribosomic RNA (rRNA)

(a) rRNA and r-proteins

The ribosomes are organelles situated in the cytoplasm, where protein synthesis takes place. These are the cells 'protein factories'. Ribosomes are also found in the mitochrondria (where the synthesis of certain mitochondrial proteins takes place).

The cell whose ribosomes have been studied most is *E. coli*.

A ribosome (70 S) of *E. coli* is formed from two sub-units:

— a large sub-unit (50 S) and
— a small sub-unit (30 S) (S stands for Svedberg — the Svedberg unit being the unit of measure for the speed of sedimentation). The coefficient of sedimentation of a particle depends not only on its mass but also on its shape and rigidity (which explains the fact that the combination of a 50 S sub-unit and a 30 S sub-unit can give a 70 S ribosome).

Each of these two sub-units consists of a mixture of:

— proteins: 'r-proteins'
— RNA: 'rRNA'

The rRNAs are the major constitutents of the ribosome (about 65%).

Much progress has been achieved in recent years, thanks to immunological techniques and electron microscopy, in understanding the structure of the ribosome.

By way of indication:

- the 30 S sub-unit contains:
  - — one 16 S rRNA 1542 nucleotides long (one molecule).
  - — 21 different r-proteins, one molecule of each. These proteins are designated by S1, S2...S21 (in the case of the r-proteins, S signifies 'small', i.e. the r-protein of a small sub-unit, whereas in the case of the r-RNAs, S signifies Svedberg unit and goes with the sedimentation constant).

rRNA and r-proteins together form a 900 000 Da (Daltons) structure.

- The 50 S sub-unit contains:
  - — one 5 S rRNA of 120 nucleotides (one molecule)
  - — one 23 S rRNA of 2904 nucleotides (one molecule)
  - — 34 r-proteins designated by L1, L2... (L for 'large', i.e. 'large' sub-unit). 30 r-proteins are present only once, and one r-protein is present four times.

The whole forms a 1.6 million Da structure.

When the small and the large sub-units come together, a track is formed between the two sub-units into which the mRNA passes.

We shall speak again later on the problem created for the cell by ribosome synthesis — in particular the synthesis of 52 different proteins (and hence the involvement of 52 genes) — without wastage.

Protein synthesis consumes a great deal of energy and the cell will adjust the synthesis of its own ribosomes to suit its needs.

In eukaryotes, ribosomes are a little larger (80 S, the two sub-units being 40 S and 60 S). Their composition is also slightly different: proteins and rRNAs are a little longer (four have been found instead of three: 18 S in a small sub-unit; 5 S, 5.8 S and 28 S in the large). This difference of ribosome constitution is of considerable significance. Certain antibiotics, those which act at ribosome level, will then have a specific action on the bacterial ribosomes without prejudice to our own ribosomes.

(b) Role of rRNAs

The precise role of rRNAs is still not known. However, they have:

- — a structural role, as we shall see, since they partly constitute the substance of a ribosome.
- — a role in facilitating fixation of other RNAs (tRNA, mRNA) on the ribosome.

2. Transfer RNA (tRNA)

(a) What is tRNA?

The tRNA molecules are so called because they will transfer, or carry, the amino acids which are found in the cytoplasm to a ribosome, during protein synthesis.

Prokaryotes

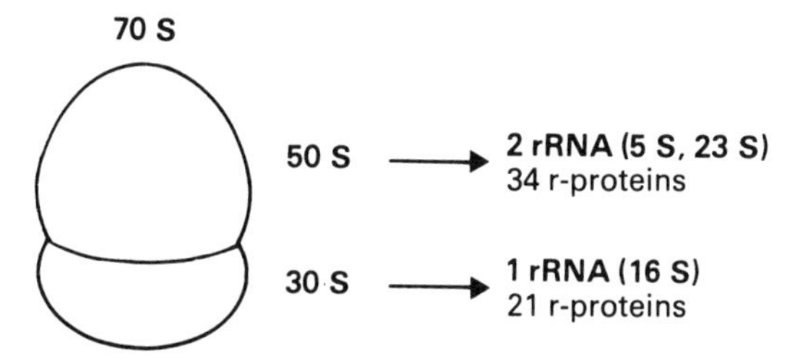

Eukaryotes

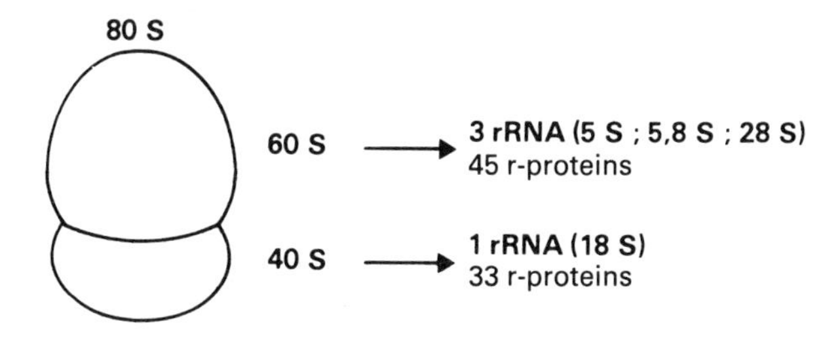

Fig. 30 — rRNAs and r-proteins.

(b) Structure of tRNA

A tRNA molecule possesses the general structure of the RNA. What are the characteristics proper to a chain of tRNA?

- Atypical nucleotides

A tRNA contains atypical nucleotides, i.e. it is unusual as regards the nature of the bases it comprises. For example, hypoxanthine is found, whose corresponding nucleotide is IMP (we shall see the significance of hypoxanthine in the paragraph where 'wobble' is studied.)

Bases such as thymine are also found, which is unusual in RNA, as well as other methylated bases.

These unusual bases are not, however, incorporated, as are the usual bases, at the moment of synthesis of tRNA. They are formed secondarily, by modification of one of the four bases (A, U, C, G) normally used during synthesis. Thus, IMP comes from the deamination of the adenine in AMP (NH_2 being replaced by OH), TMP comes from the methylation of UMP. (We shall speak later of these so-called 'post-transcriptional' modifications.

- Spatial form of tRNAs

The tRNA chain (comprising about a hundred nucleotides) folds up to give the general appearance of a clover leaf. The explanation is:

— the trefoil branches are formed by hydrogen linkages between complementary

bases; these are paired regions,
— The loops on the other hand are formed from non-paired nucleotides; atypical nucleotides are found there amongst others.

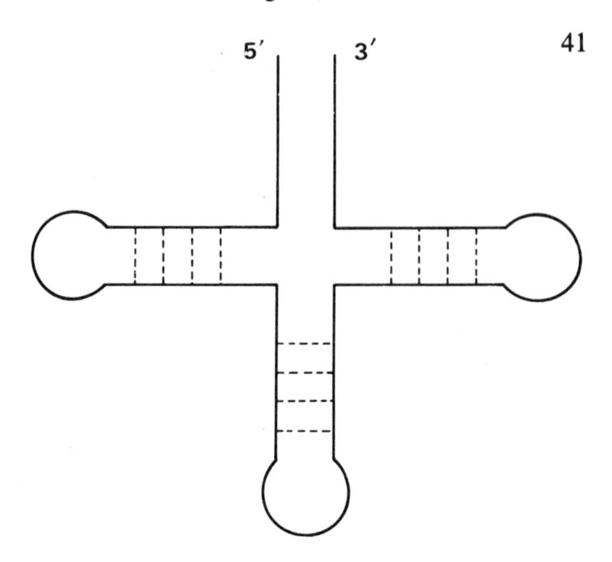

Fig. 31 — Representation of a tRNA molecule.

In space, in fact, this molecule folds to give a shape resembling an L (this is because a double helix forms in two places of the tRNA and these two helicoid segments lie perpendicular to each other). This three-dimensional structure of tRNA has become known thanks to crystallographic studies with X-rays.

- Diagrammatic representation of tRNA

To simplify this we could schematise a tRNA molecule as in Fig. 32.

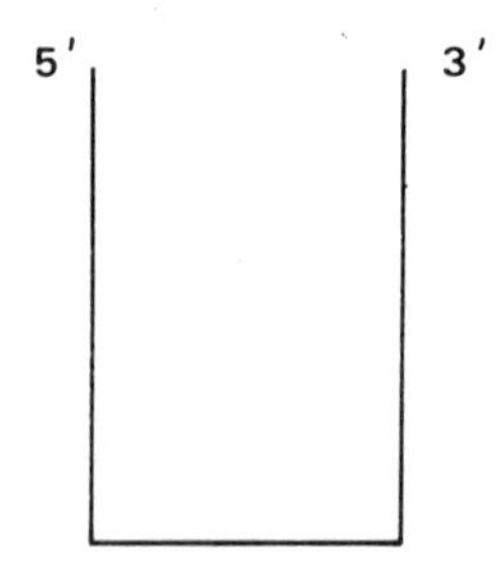

Fig. 32 — Highly schematised representation of a tRNA molecule.

In fact we shall see that the tRNA associates with an mRNA molecule in an antiparallel way. To simplify this we shall from now on adopt the convention of always writing mRNA (in the 5′-3′ direction) from left to right. The tRNA will then be read (in the 5′-3′ direction) from right to left, as represented in Fig. 33.

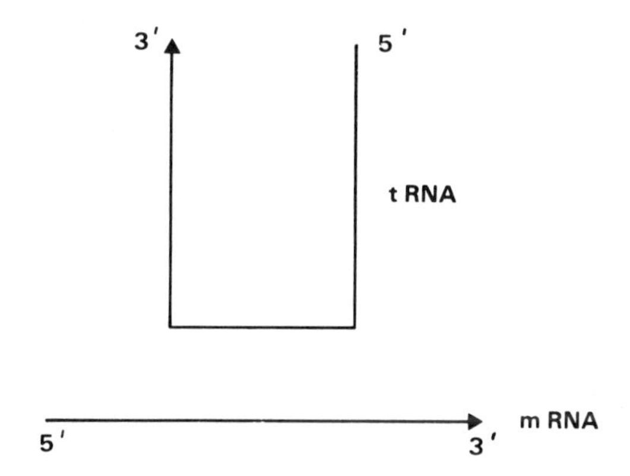

Fig. 33 — tRNA and mRNA are antiparallel.

- Two sites are important in a tRNA molecule

— The 3′OH end

The extremity of all tRNA molecules ends with the three following nucleotides: CCA. Do not forget that 'CCA' in fact symbolises three nucleotides and not just three bases. Each tRNA then ends with the three nucleotides: CMP, CMP, AMP.

It is to this CCA end that the amino acid (aa) to be transported will be fixed.

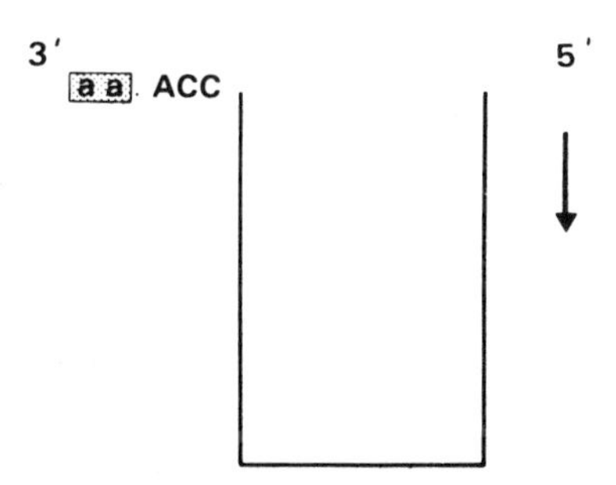

Fig. 34 — The amino acid is carried to end 3′ of a tRNA.

— The anticodon

A group of three nucleotides (a triplet) situated on a loop of tRNA is called an anticodon. This anticodon plays a very important role since it will recognise the codon, a group of three nucleotides located on the mRNA.

This codon-anticodon matching occurs through weak bonds (hydrogen bonds), is antiparallel and shows complementarity between the bases of the codon and the anticodon.

In representations employing codons and anticodons it is easier, as we have mentioned already, to write the codon in the 5′–3′ direction and the anticodon in the opposite direction. If, in commenting on a figure, one is led to write an anticodon in the text in the reverse direction from 5′–3′, it would be best to specify this by writing the numbers 3′ and 5′ on top.

$$\overset{3'\leftarrow 5'}{\text{UAC}} \quad \text{is equivalent to:} \quad \overset{5'\rightarrow 3'}{\text{CAU}}$$

(c) How does tRNA–amino acid bonding take place?

- Type of bonding

tRNA carries its amino acid to a ribosome 'after having latched onto it' by means of a covalent bond. This is an ester bond. This bond is effected by the elimination of a molecule of water between an acid function carried here by the amino acid and an alcohol function carried here by the tRNA (as we shall see, the alcohol function of the tRNA is an OH group of the ribose situated on the last tRNA nucleotide, that is to say the nucleotide AMP.

- Intervention of aminoacyl~tRNA synthetase and energy supply

This ester bond does not occur simply through reaction between an amino acid and a tRNA molecule which would both be free in the cytoplasm. In fact it requires:

— The intervention of an enzyme called aminoacyl~tRNA synthetase. It is onto this enzyme that the amino acid and the corresponding tRNA first become fixed to finally give aminoacyl~tRNA;

— An energy supply (a little more than 8000 calories). This energy is carried by the ATP which is a molecule possessing two bonds of high energy potential called 'energy-rich bonds' in energy represented by the symbol~. As we shall see, two energy rich bonds will be needed to form a molecule of aminoacyl~tRNA.

- The different stages

The esterification reaction takes place in two stages.

— First stage: latching of the amino acid onto the AMP.

This stage is called activation of the amino acid. It ends with the production of the aminoacyl~AMP (aa~AMP).

$$\text{aa} + \text{ATP} \rightarrow \text{aa} \sim \text{AMP} + \text{Ⓟ}\sim\text{Ⓟ} \;(\llcorner 2\ \text{Pi})$$

$$\underset{\overset{|}{\text{CH}}}{\text{H}_2\text{N}\ \text{COOH}} + \text{HO (ATP)} \rightarrow \underset{\overset{|}{\text{CH}}}{\text{H}_2\text{N}\ \ \text{CO} \sim \text{O AMP}} + 2\ \text{Pi}$$

↑ (anhydride)

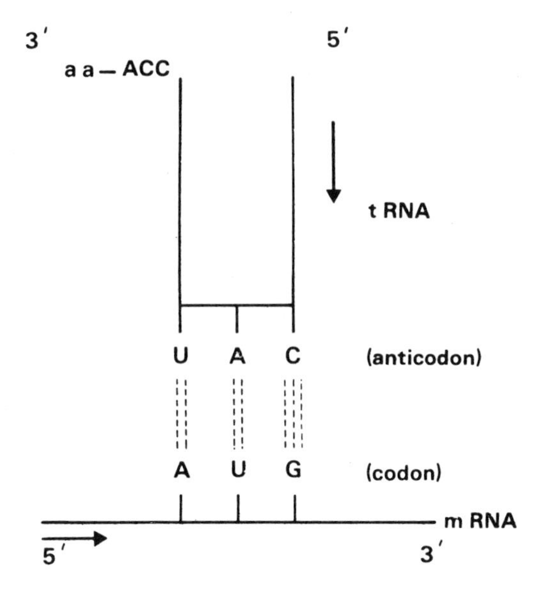

Fig. 35 — The anticodon (on the tRNA) is complementary and antiparallel to the codon (on the mRNA).

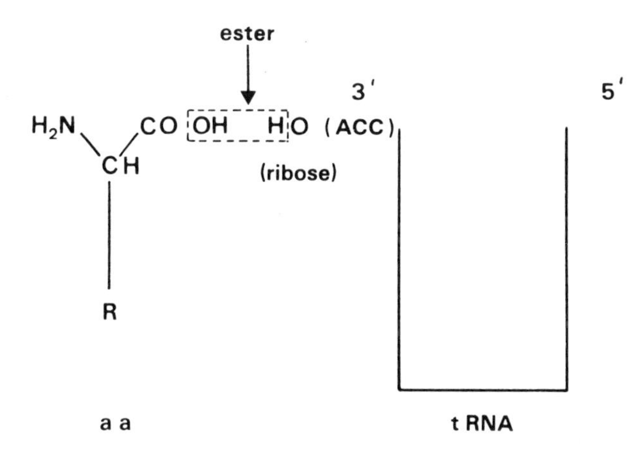

Fig. 36 — Bond between amino acid (aa) and tRNA.

The bond between the amino acid and the AMP is an acid anhydride bond. It is produced by the elimination of a molecule of water from between the acid function of the amino acid and an acid function of the phosphoric acid of the AMP nucleotide. An acid anhydride bond is rich in energy.

The pyrophosphate formed in this area is hydrolysed by a pyrophosphatase. Two molecules of mineral phosphates (or 'P_i', i for inorganic) then form.

— second stage: transfer of the amino acid (from aa~AMP) onto the tRNA

This ends with the formation of aminoacyl~tRNA (aa~tRNA)

aa~AMP + tRNA → aa~tRNA + AMP

H_2N—CH(—)—CO~AMP (anhydride) + HO (tRNA) (ribose) → H_2N—CH(—)—CO~tRNA (ester) + AMP

There is now an ester bond between the amino acid and the alcohol function of the ribose carried by the last nucleotide of the tRNA, i.e. carried by the AMP. The OH group of the ribose participating in this ester bond may be either OH at 2′, or OH at 3′. Once the bond is constituted, the aminoacyl grouping can 'jump' between positions 2′ and 3′.

The final balance of these two equations is then the following:

aa + tRNA + ATP → aa~tRNA + AMP + 2 Pi

(ATP: 2~) (aa~tRNA: 1~)

Let us now look at the three formulae relating to the products entering into these reactions: ATP, aa~AMP (or activated amino acid), aa~tRNA.

To sum up: in both the first and second stages amino acid is joined to one AMP by its acid function. But:

— in the first stage (where there is formation of aminoacyl~AMP), there is an anhydride bond between the amino acid and AMP;
— in the second stage (where there is formation of the 'aminoacyl~tRNA'), it is a case of an ester bond between amino acid and the last AMP of the tRNA.

Remarks on

- The ester bond in aa~tRNA (rich in energy)

This bond is very curious as regards energy. In fact it is an energy-rich bond although as a rule, an ester linkage is not an energy-rich combination. Here, the energy which was contained in the acid anhydride linkage (aa~AMP) is transferred to the ester

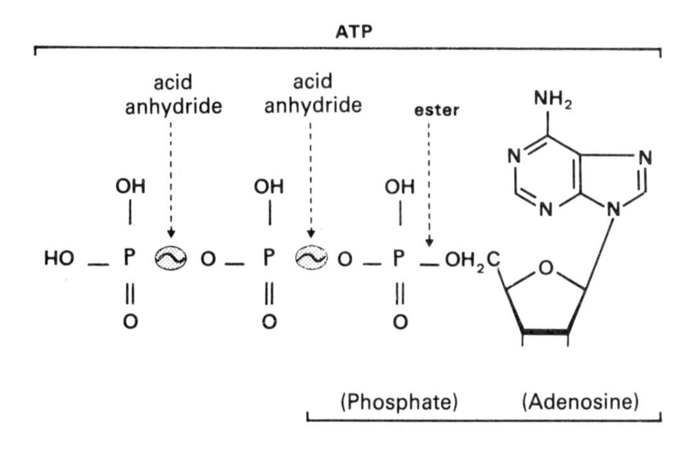

Fig. 37 — Formula for ATP.

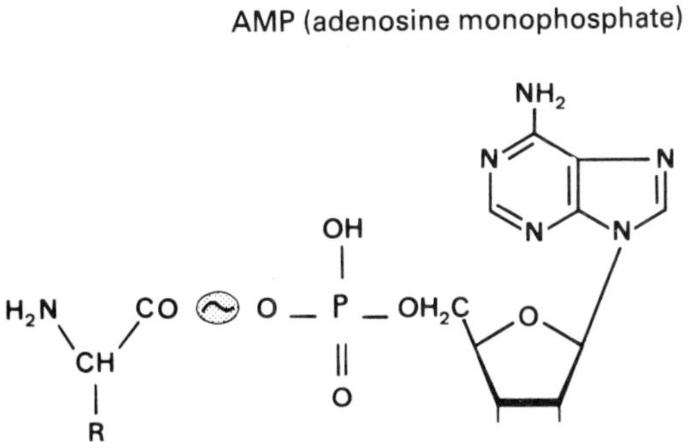

AMP (adenosine monophosphate)

Fig. 38 — General formula for an amminoacyl~AMP.

bond (aa~tRNA).

Thus, the various tRNAs will not only carry the amino acid but also the energy needed to attach the amino acids to each other at the moment of protein synthesis. (This bond between two amino acids will be an amide bond called a 'peptide bond').

In fact, as we shall see further on, the energy-rich bonding of an aa~tRNA will be utilised, not to fix this amino acid to the peptide chain in formation, but to attach the next amino acid in the form of another aa~tRNA.

Two energy-rich bonds originating from a molecule of ATP have had to be used to form this ester bond of aa~tRNA:

— the first energy-rich bond is used to give aa~AMP (then aa~tRNA by transference)
— the second energy-rich bond used is due to pyrophosphate hydrolysis.

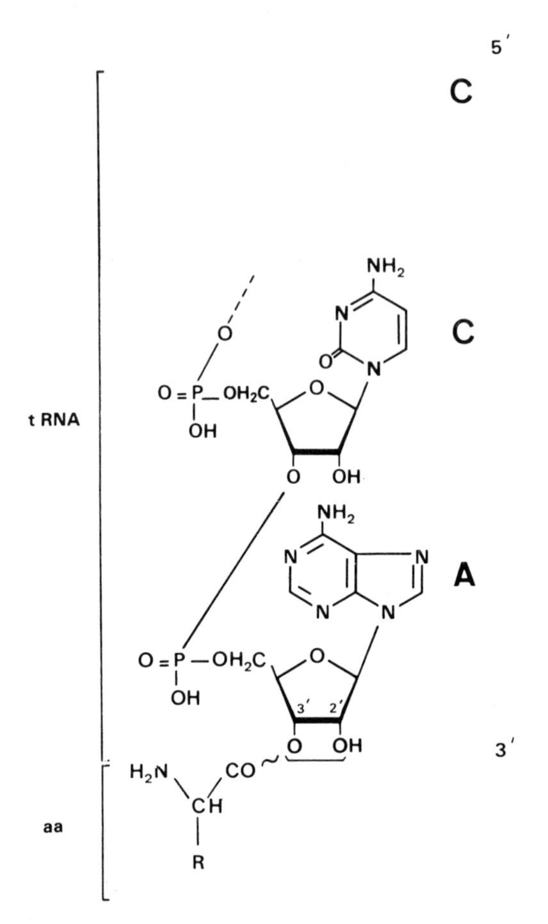

Fig. 39 — General formula for an aminoacyl~tRNA.

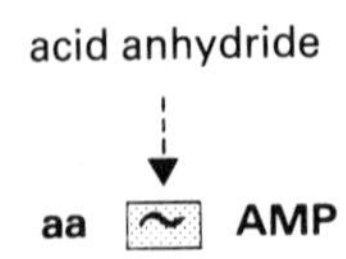

Fig. 40 — Acid anhydride bond betwen amino acid and AMP.

The two energy-rich combinations have enabled an energy-rich ester bond to be formed, and, thanks to the pyrophosphate hydrolysis, the reaction giving activated amino acid is irreversible.

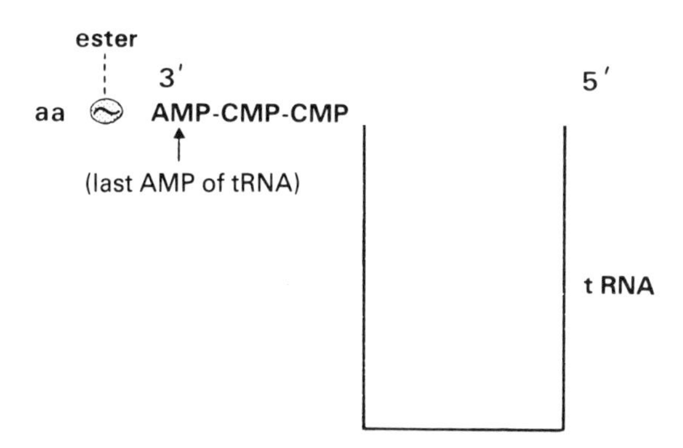

Fig. 41 — Diagrammatic formula of the 3′ end of an aminoacyl~tRNA.

Aminoacyl~tRNA synthetase (editing enzyme)

One may ask oneself how a tRNA molecule recognises its amino acid when the various tRNAs all terminate in the same way with 'CCA'.

We now know that it is not the tRNA which directly fixes its amino acid, but an enzyme 'amino acyl~tRNA synthetase'. This enzyme recognises at the same time:

— the 'good' tRNA, and also
— the corresponding 'good' amino acid.

This enzyme successively brings about:

— activation of the amino acid at aa~AMP,
— the linkage of the activated amino acid to tRNA.

The aminoacyl~tRNA synthetase therefore has as its substrates the amino acid, ATP and tRNA, and all the reactions previously mentioned will be able to take place without the intermediaries (activated amino acids) leaving the enzyme.

The aminoacyl~tRNA synthetases are very specific enzymes. This stage, where aminoacyl~tRNAs are formed is very important. It is essential that no errors occur during this phase. However, although two different tRNAs are easily recognised by the enzyme (by virtue of the different sequence of bases, but above all of the different tertiary structure of these two tRNAs), it is not always the same for the amino acids. A valine for example, is very similar to an isoleucine. If a Val is activated by mistake, the Val~AMP formed will fortunately not be fixed to the tRNA of the Ile. The enzyme (the isoleucyl~tRNA synthetase) is capable of correcting its own mistakes! In fact, it will then hydrolyse this inadvertently formed Val~AMP (in a hydrolytic site different from the site of synthesis).

It is said of such enzymes that they have an 'editing function'. In fact they are capable, like an editor before the final printing, of guaranteeing a 'proof reading' followed by an 'error correction'. (This remarkable situation where an enzyme is capable of correcting its own errors will be seen again later with other enzymes which also have to be highly specific, the DNA polymerases.)

It is important to understand that it is the tRNA which, through its anticodon, determines which amino acid will be incorporated into the peptide chain. This can be demonstrated by a very ingenious experiment.

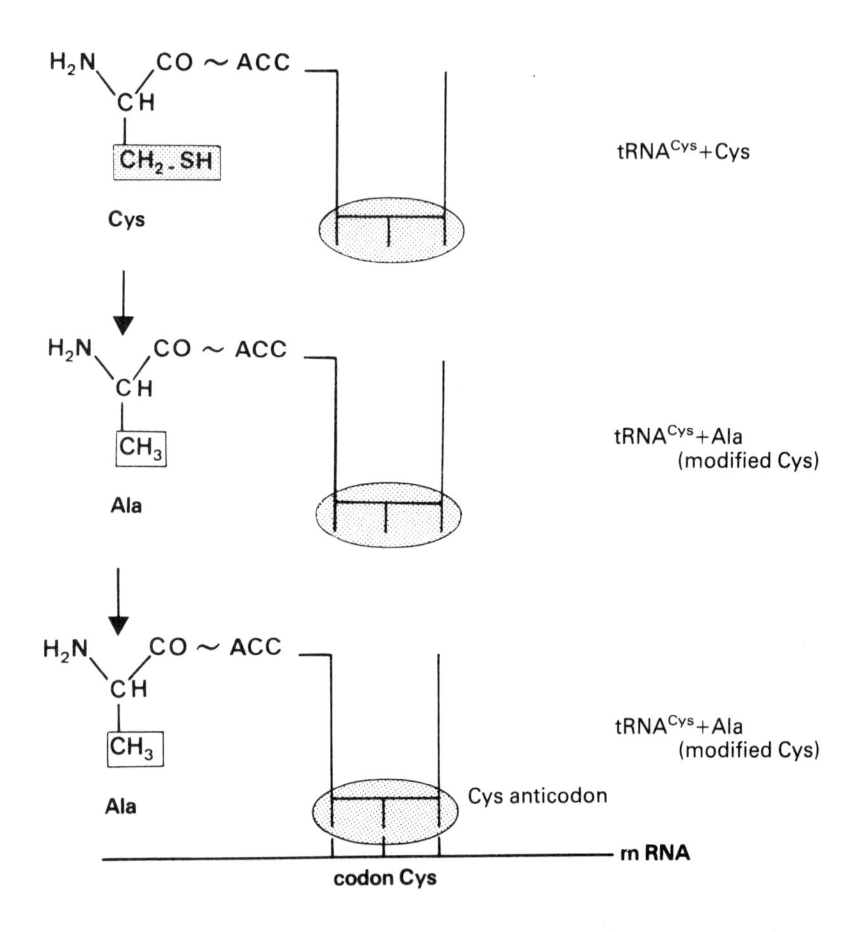

Fig. 42 — Experiment demonstrating that a modification at the level of the amino acid carried by the tRNA does not modify codon–anticodon recognition.

Take cysteinyl~tRNA (the tRNA carrying the Cys). Cys is transformed into Ala by a chemical reaction. It can be shown that it is the Ala which is then incorporated into the peptide chain in response to a Cys codon! In fact, the anticodon of tRNA has not itself been modified. This is why it continues to recognise the codon corresponding to the Cys. It is therefore the amino acid (whatever it may be) carried by tRNA, whose anticodon matches the Cys codon which will finally be incorporated.

Given that there are 61 different codons (coding for 20 amino acids) one should have believed that 61 different tRNA codons existed. It will be seen later (section on 'wobble') that this number is less than 61. In any case it is more than 20. This means that an amino acid can be transported by several tRNAs specific to that amino acid — but differing in their anticodons — whereas one type of tRNA can only transport a single amino acid.

- Queuine

A particular base has been found in four tRNAs ($tRNA^{Asn}$, $tRNA^{Asp}$, $tRNA^{His}$, $tRNA^{Tyr}$) in eubacteria and in the eukaryotes (except yeast): queuine (Q) (pronounced 'kwee-yoo-een'). This base has a formula similar to guanine, but it has only one nitrogen atom (instead of two) in the pentagonal ring. In addition this pentagonal ring carries an aminomethyl dihydroxycyclopentene substituent. The corresponding nucleoside is queuosine. When queuine is present on a tRNA it occupies position 1 of the anticodon.

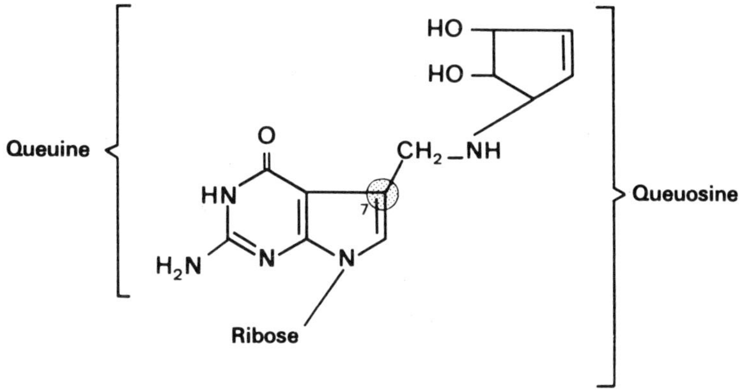

Fig. 43 — Formula for queuosine.

Mammals (unlike bacteria and plants) are unable to synthesise queuine, which they obtain from an exogenous source. Guanine (situated on the anticodon) is replaced by queuine. The enzyme intervening in this reaction is a tRNA guanine transglycosylase.

In mammals, Q may be implicated in cellular differentiation and neoplastic transformation. The percentage of tRNA having Q instead of G changes in the course of development. Thus the percentage of tRNA with guanine is higher in fetal tissue and in certain varieties of animal tumours.

3. Messenger RNA (mRNA)

(a) Definition

This type of RNA is called 'messenger' because it carries part of the genetic information contained in DNA to the ribosome where it will carry out protein synthesis, as previously discussed.

(b) Life-span

The life-span of mRNAs is very short. In this respect it differs from the much longer life of tRNA (since each tRNA carries its amino acid to the ribosome, liberates this amino acid and can thus be re-used many times).

mRNAs renew themselves very rapidly; they are produced and degraded rapidly. They only last (like roses) as long as the message takes — one mRNA, however, can be read may times in the ribosome. In bacteria, the life span of a mRNA is just a few minutes. In eukaryotes, mRNAs are more stable (several hours to several days).

(c) Structure

Like other RNAs, mRNA is formed from a single nucleotide chain comprising the same kind of bases: A, U, C, G. Where then are these messages which mRNA is reputed to contain? We shall see that it is the nucleotide — or more exactly the sequence of bases — which in fact constitutes a message, in the form of a code; until recently a 'secret' code. Each group of three nucleotides on the mRNA forms a codon. Each codon will code for a quite specific amino acid. The decoding of the message carried by the mRNA will be performed at ribosome level. Thus, codon AUG will signify 'attachment of a methionine during protein synthesis'; the UUU codon will signify 'attachment of the phenylalanine', etc.

2

Protein synthesis

Protein synthesis comprises two important stages:

— transcription,
— translation.

1. TRANSCRIPTION

A. Definition

Transcription is the mechanism by which mRNA is synthesised.

mRNA is formed by copying a portion of DNA and it is important to understand that:

— not all DNA is transcribed, but only certain portions of the DNA. The transcribed DNA sequences are called 'genes' (the genes and non-coding sequences of an individual are collectively called the 'genome').
— Only one of the two DNA strands is copied, but it is not always the same strand of DNA which is copied the whole length of a DNA molecule: for certain genes it would be one strand, for other genes it would be the other strand.

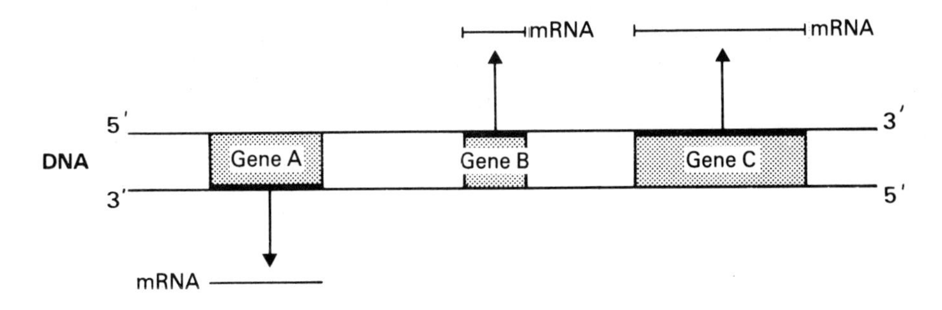

Fig. 44 — Transcription of DNA: only one of the two DNA strands is transcribed (it is not always the same strand which is transcribed, depending on different genes).

B. Characteristics

Synthesis of a mRNA is achieved:

— in the $5' \rightarrow 3'$ direction
— in an antiparallel way (with respect to the strand of DNA transcribed) and
— complementarity

C. Elements needed for transcription

To synthesise mRNA *in vivo,* the following must be present in particular:

1. Nucleotides

— These nucleotides must contain bases A, G, C, U. (They differ from DNA nucleotides in that U replaces T, and the ribose replaces Deoxyribose).
(a) The four types of nucleotide must be activated, not in the form of monophosphate nucleosides, but as in the triphosphate form (ATP, GTP, CTP, UTP).

2. An enzyme: RNA polymerase

RNA polymerase is the enzyme which enables nucleotides to join with each other to form the polymer which is mRNA.

3. A DNA model

mRNA being a copy (complementary and antiparallel to DNA) it is essential, in manufacturing a mRNA, to have a DNA strand available as a model. As the RNA polymerase could not function without DNA, it is said to be 'DNA-dependent'.

D. The different stages of transcription

Transcription is a selective phenomenon: not all DNA is transcribed. The problem consists of knowing how the sequences to be transcribed will be recognised by the RNA polymerase. In fact the questions posed are:

— What will indicate that an area of DNA should start being transcribed?
— Similarly, what will indicate that transcription should cease?

1. The start of transcription

The signal which starts transcription is the 'promoter'. The promoter is an area of DNA comprising about 40 pairs of nucleotides situated just ahead of the start of the area where transcription will begin. By convention, the first nucleotide is called +1.

The nucleotide sequences of more than 100 promoters found in *E. coli* have now been described. These promoters contain highly conserved sequences. (It is also said that these sequences present a homology). In particular there are two short sequences of six pairs of nucleotides separated from each other by about 25 pairs of nucleotides.

(a) Sequence '−35'

One of the preserved sequences is situated approximately 35 pairs of nucleotides upstream of the point of departure of the transcription (from −35 to −30 in the example below). Three bases are highly preserved ($>75\%$): TTG . . .

(b) Sequence '−10'

The second sequence is situated about 10 pairs of nucleotides upstream of the point of departure of transcription (from −12 to −7 in the example below). Three bases here are equally highly preserved: TA . . . T

This sequence '−10' is also called 'the Pribnow Box' and can be considered as an area advertising for RNA polymerase. The reverse count starts: −6, −5, −4, −3, −2, −1 and +1! Transcription starts . . .

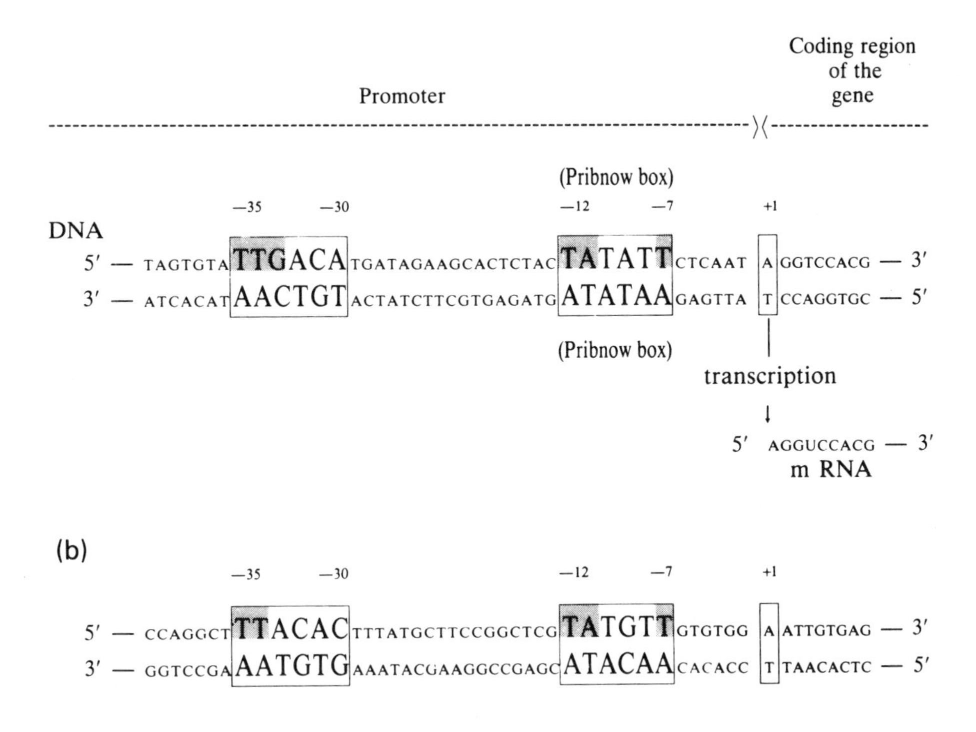

Fig. 45 — The promoter. (a) Sequence of one of the promoters of *E. coli*, (b) Sequence of the promoter of the *lac* operon of *E. coli*.

The signals which initiate transcription are less well known in the eukaryotes where several RNA polymerases (I, II, III) exist. RNA polymerase I transcribes the genes for certain rRNAs, whereas III transcribes the genes for the tRNAs in particular (and other rRNAs). Only II transcribes the genes which will be expressed in proteins. RNA polymerase II (the one which synthesises the mRNAs) recognises:

- a sequence situated at about −130 b.p. upstream of the point of departure of transcription ('CAAT box'),
- a sequence situated at about −30, rich in AT bases, called the 'TATA box'. It is also called 'Goldberg Hogness box' (from the name of the researchers who

discovered it). This sequence is to some extent the equivalent of the Pribnow box of the prokaryotes.

Mutations at the TATA box level result in a 50–70% inhibition of transcription. (However, transcription is not affected by a mutation taking place at the point of departure of transcription.)

2. *Transcription proper*

(a) *Formation of ester bonds between nucleotides*

Transcription then consists of fusion of nucleotides to each other by an ester bond. The attachment of each new nucleotide will take place in the $5' \rightarrow 3'$ direction.

The energy needed to form the ester bonds between the different nucleotides is supplied by the nucleotides themselves, which have to be present, as we have seen, in the form of triphosphate nucleosides: ATP, GTP, UTP, CTP.

The triphosphate nucleosides thus supply at the same time:

— the monophosphate nucleoside
— the energy needed to bind each nucleotide to the preceding one.

It should be noted that only the first nucleotide of the mRNA preserves its triphosphate grouping.

(b) *Progression of transcription*

It is important to understand that RNA synthesis takes place in the $5' \rightarrow 3'$ direction.

RNA polymerase can be visualised as a cursor moving about on a display screen. The two DNA strands drift apart through the rupture of hydrogen bonds. Transcription begins (the copying of one strand of DNA). But gradually, as the displacement of RNA polymerase on the DNA proceeds, mRNA is produced and then becomes detached, having appeared temporarily on the transcribed strand of DNA. The hydrogen bonds between the two strands of DNA reform behind the RNA polymerase. The two strands of DNA reassume their helicoid shape.

(This transcription progression could also be visualised as an 'eye' advancing along the DNA (see Fig. 47).)

3. *End of transcription*

In prokaryotes, there are two signals which terminate transcription.

- An area of imperfect symmetry

In order to understand what an area of imperfect symmetry or 'imperfect palindrome' is, let us go back to a literary example (if it can be so called):

'no misses order red roses Simon' is an example of a perfect palindrome

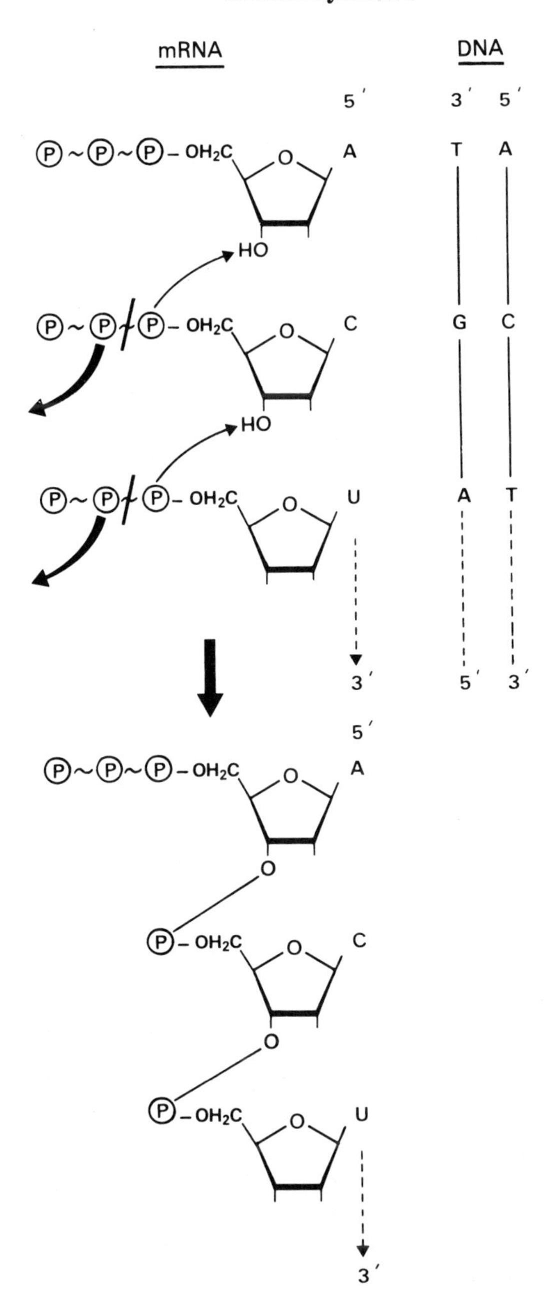

Fig. 46 — Transcription of DNA: example showing the transcription of the first three nucleotides.

'no misses saw dead roses Simon' is an example of an imperfect palindrome.
It is easy to see that in the second example there is an interruption of the symmetry. It is the same with the signals which terminate transcription.

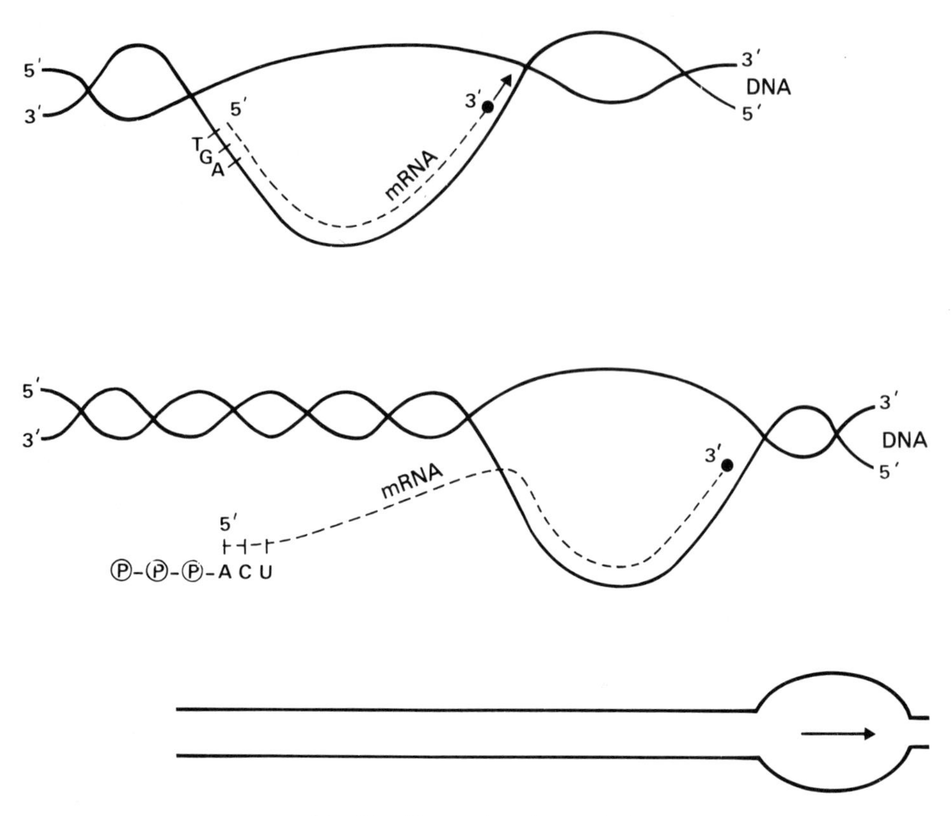

Fig. 47 — Progression of transcription.

- An area rich in AT

The area of imperfect symmetry is immediately followed by a short sequence rich in A and T bases. (The strand of DNA which is to be transcribed contains, at this stage, a sequence of several As). It is thus a more relaxed area (two hydrogen bonds between the A and T bases instead of three between bases C and G), from which the RNA polymerase can easily emerge.

Transcription of the area of imperfect symmetry will give a sequence of mRNA capable of self-complementarity. The mRNA molecule will end with a loop followed by several Us (coming from transcription of the sequence with several As). It is this horseshoe helicoidal loop which is responsible for stopping transcription (rather than the sequence by itself).

With eukaryotes, the imperfect palindrome is located further upstream, on the other hand the signal to terminate the gene (and not to end transcription) would be:

AATAAA (the sequence read on the strand of DNA not transcribed).

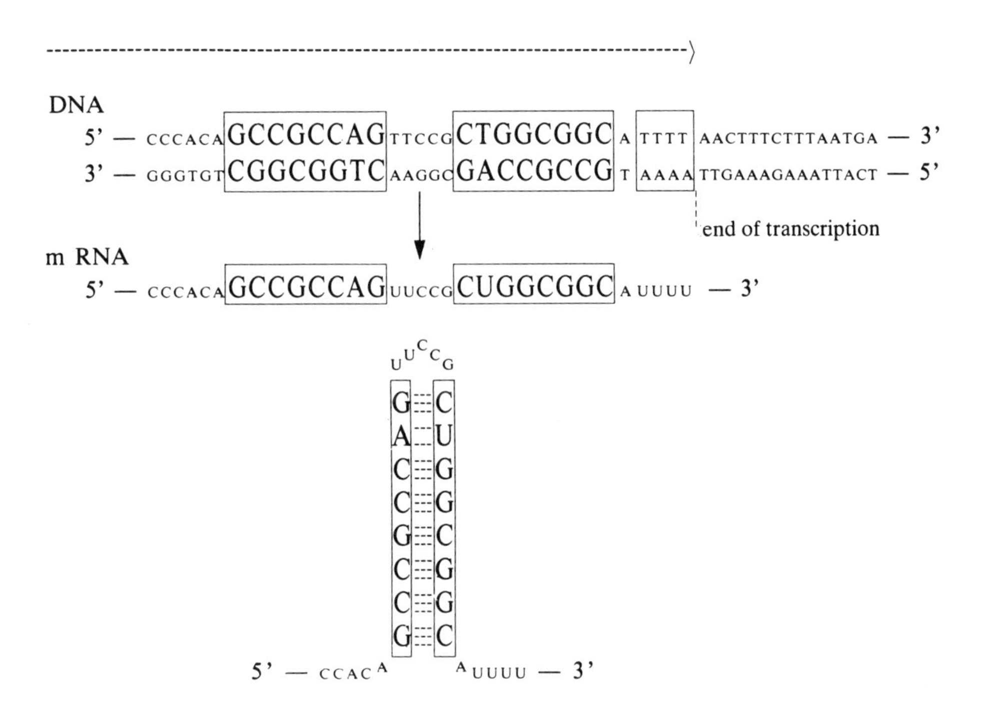

Fig. 48 — Sequence at the end of transcription in prokaryotes.

The RNA polymerase recognises this signal on the DNA but continues to transcribe beyond it. However, the transcripts are curtailed later and will terminate with the signal AAUAAA followed by 10–15 nucleotides (before receiving a poly-A tail).

4. Remarks on transcription

(a) Which strand of DNA is copied by the RNA polymerase?

As already mentioned, RNA is formed from the 5′ end towards the 3′ end (each new nucleotide will then be placed at 3′), and in an antiparallel manner in relation to the strand of DNA transcribed. It is the direction of movement of the RNA polymerase along the DNA which determines which strand will be copied. In Fig. 49 an RNA polymerase molecule which goes from left to right will transcribe the lower strand (after having encountered the gene's promoter). Thus, we have an antiparallel transcript. Conversely, an RNA polymerase molecule travelling along the DNA from right to left will transcribe the upper strand of DNA (after having met the corresponding gene promoter).

It should be understood that if the mRNA is complementary and antiparallel in relation to the strand of DNA transcribed, it is by contrast strictly identical to the second strand of DNA (the one which is not transcribed) both in polarity and in base sequence (with the exception, however, of the T bases of the DNA replaced by U in the mRNA). For example:

```
5' ... A G C T T A A C G C G T A ... 3'   Strand of DNA not transcribed
3' ... T C G A A T T G C G C A T ... 5'   Strand of DNA transcribed
                 |
           Transcription
                 ↓
5' ... A G C U U A A C G C G U A ... 3'   mRNA
```

(b) What is the role of the gyrase (bacterial) after the initiation of transcription?

Let us first of all discuss the 'looped domain'. A 'looped domain' is a loop of DNA which is somehow 'clamped' (by the proteins) at the loop's neck level. This loop of DNA represents a segment of DNA, the extremities of which are relatively fixed. Thus, the two strands of DNA are not free, since there is a tension affecting the molecule, turning one around the other. Numerous 'looped domains' exist in the DNA of *E. coli*.

At the start of transcription, each time the RNA polymerase begins its synthesis of mRNA, one double-helix turn has to take place (along a length of about 10 b.p.). The RNA polymerase is, unlike the topoisomerases, incapable of cutting the strands of DNA. The unrolling (negative overtwisting) of 10 b.p. must therefore be necessarily compensated for by a positive overtwisting in the immediate proximity. This would create a tension unfavourable to the series of reactions. The role of the gyrase is to prevent this kind of difficulty. It serves to 'absorb', to 'suck up' every positive supertwist which is formed. In fact, as previously discussed, the gyrase forms negative supertwists (beforehand) so well that the majority of bacterial DNA is retained in negatively overtwisted form. Thus, in the place where the action of the RNA polymerase produces the formation of a positive supertwist, it will in fact bring about the suppression of a negative supertwist.

The diagrams below show how it is easier to suppress a negative supertwist than to create a positive supertwist (certainly when the DNA ends are fixed).

In the eukaryotes, the mechanism by which the two strands of DNA drift apart at the start of transcription is not yet precisely known. It is all obviously much more complicated since the DNA is packaged with the histones in the nucleosomes as we shall see later.

(c) Amplification of the information contained in DNA

A gene then is a scrap of DNA, one of the two strands of which will be transcribed. Starting with a single gene, and thus a single segment of DNA which serves as a model, transcription will give rise to numerous specimens of mRNA, all containing the same information. It is the principle of an 'original' of DNA and of several

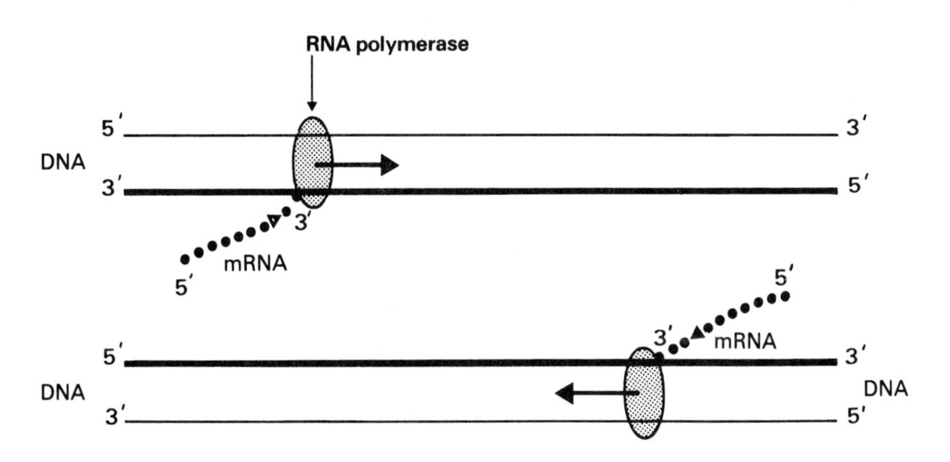

Fig. 49 — Progression of the RNA polymerase.

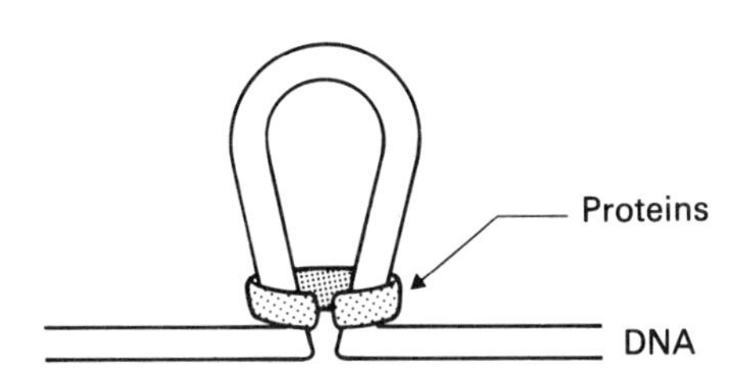

Fig. 50 — A 'looped domain'.

'photocopies' of RNA. In fact, given the complementarity, it would be fairer to use the image of a 'negative' DNA and of several 'positive' mRNAs. One gene will be able to give, for example, 100 mRNAs. Thus at this level a very important amplification phenomenon exists, which is similarly found during translation — so much so that a single gene can be the source of several thousand identical proteins.

It is not necessary, therefore for DNA to contain thousands of identical copies of the same gene.

E. What are the products of transcription?

Transcription consists, as we have seen, of producing a RNA complementary to a strand of DNA, but transcription does not only produce mRNAs.

The different products of transcription can in fact be:

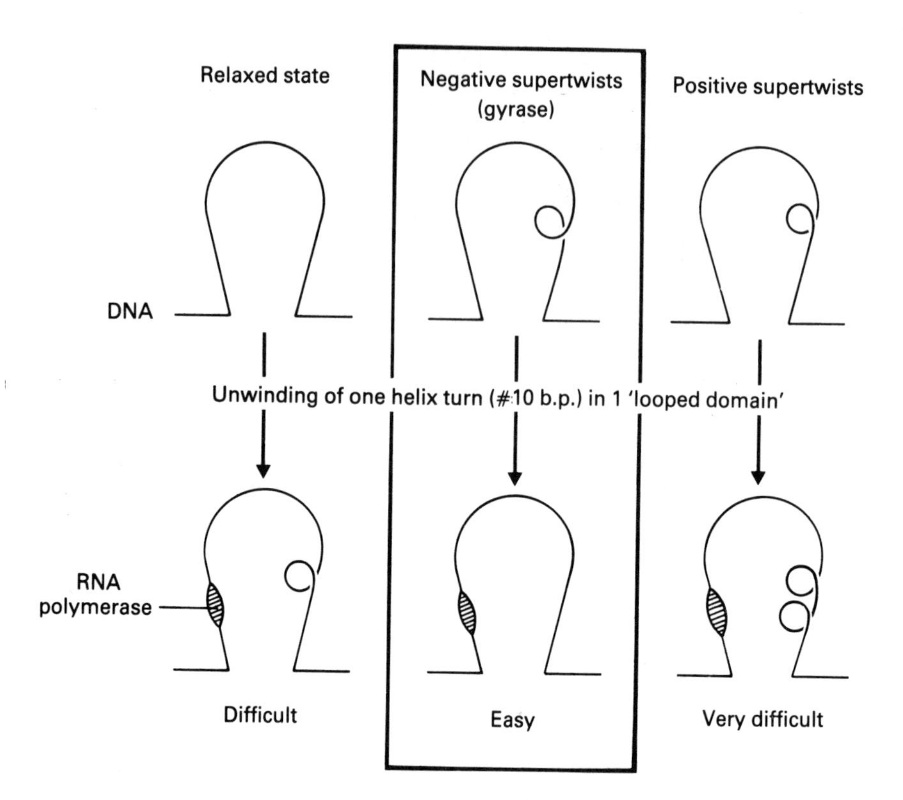

Fig. 51 — Diagram showing how the start of transcription is facilitated if the DNA contains negative supertwists. (Alberts B *et al. Molecular biology of the cell.* Garland Publishing, New York, p. 447, Figs 8–9).

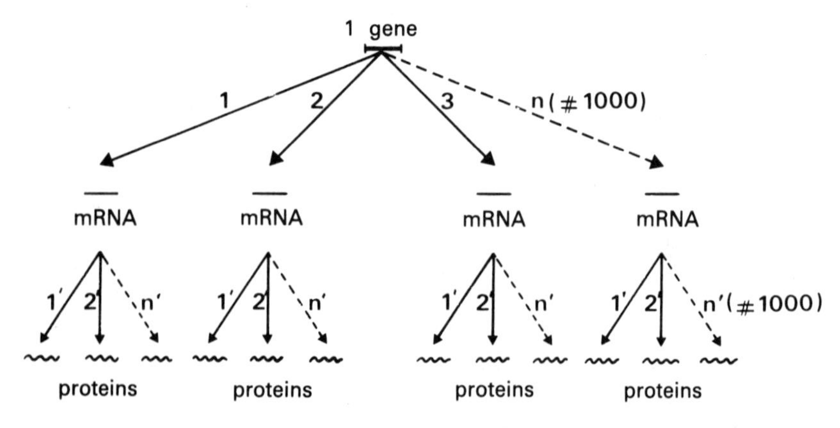

Fig. 52 — Amplification produced by transcription and translation.

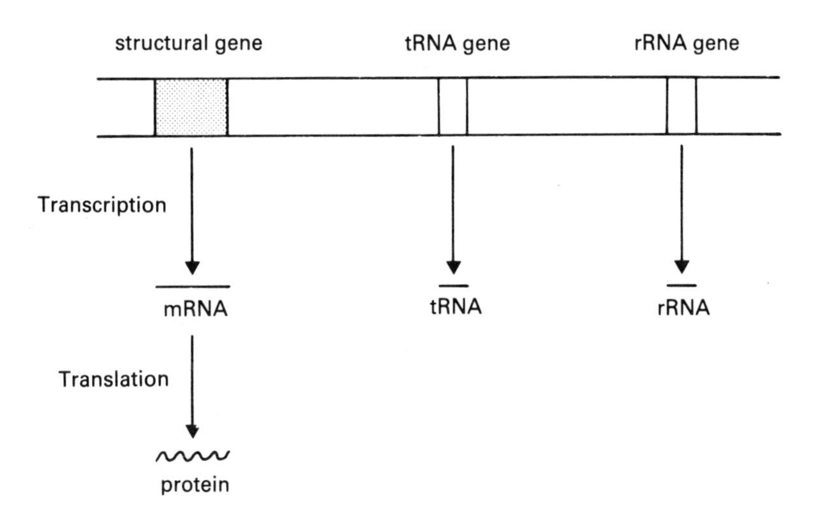

Fig. 53 — The different products of transcription (mRNA, tRNA, rRNA).

1. mRNAs

We shall see this later. The mRNAs formed have a short life-span, because they are simply destined to be translated into proteins at ribosome level.

In this case transcription is followed by translation.

2. tRNAs and rRNAs

The tRNAs and the rRNAs are synthesised by transcription of a gene. (In fact transcription gives a precursor which will then be cleaved). These RNAs are stable, unlike the mRNAs.

The tRNAs and rRNAs resulting from the transcription of segments of DNA will not be translated into proteins at ribosome level. Transcription here is not followed by transalation.

In the case of tRNAs and rRNAs, the process of amplification cannot be fully accomplished, since there will be no transcription–translation stage. However, the cell needs very many ribosomes, and thus a lot of rRNA. As opposed to genes of a protein where there is generally, as we have seen, only a single model of a gene to manufacture thousands of identical proteins, here there are many identical copies of tRNA and rRNA genes.

F. Post-transcriptional modifications

1. In prokaryotes

(a) rRNA

As we are about to show, the transcription of a rRNA gene gives a precursor which will have to be modified in order to give the final rRNA.

(b) tRNA
In the same way, a tRNA gene gives a precursor which will undergo:

— cleavages
— additions (e.g. the three nucleotides CCA will be added to the 3′ end)
— modifications such as:
 methylation (where U will give T for example),
 deamination (where AMP will give IMP).

(c) mRNA
There are practically no modifications of mRNA. However, in prokaryotes, translation starts (at 5′ of mRNA) even before transcription (at 3′ of mRNA) has ended. There is then little possibility of modifying the transcripts of RNA before their translation. (Remember, that in prokaryotes mRNA is formed directly in the cytoplasm; there are no separate nucleus–cytoplasm compartments.)

2. *In eukaryotes*

(a) rRNA and tRNA
The types of post-transcriptional modification observed in eukaryotes for the formation of rRNA and tRNA are comparable to those of prokaryotes.

(b) mRNA
The modifications carried out after transcription of DNA are very important here, as we shall see in the following section. They take place in the nucleus before the translation stage.

G. Transcription in eukaryotes

1. Structure of DNA eukaryote: exons and introns
In eukaryotes, it is said that the gene is 'discontinuous' or that the message is 'fragmented'. In fact, in the higher eukaryotes a gene comprises:

— two exons which contain the information and which will be expressed (by being translated into proteins),
— introns (or intercalary sequences) which are then interposed into the middle of the part containing the information. They will be transcribed but not translated. The role of these DNA sequences, which are apparently not expressed, is still not actually known. They can be pictured as advertising slots which interrupt a film— slots which will be eliminated when a copy (more precisely a negative) of this film is presented.

Mnemonic:
 The 'exons' are DNA sequences which are expressed.
 The 'introns' are interposed, intercalary sequences; these are intruders which interrupt the gene, and which will be untranslated).
 In fact this definition is a little oversimplified. Exons are expressed in the majority of cases, but not always. In effect, example of exons are known which are transcribed

but not translated (e.g. exon no. 1 of the apolipoprotein A-II). In this case, obviously, the transcript of the exon will not be excised as the transcript of an intron would be.

2. *The different phases of transcription*

Here is the model presently proposed for transcription in the higher eukaryotes:

(a) *Integral transcription of exons and introns*

The completeness of the gene, i.e. that both exons and introns are transcribed to give a long molecule of RNA, called the precursor of the mRNA (pre-mRNA) or the primary transcript. This phase clearly takes place in the nucleus. Modification will then be produced.

(b) *Addition of the 'cap' at the 5′ end*

The cap is a GMP having a methyl grouping on N_7 (which carries a positive charge). It is bound to the first nucleotide of the primary transcript by an acid anhydride bond. Remember that the first nucleotide of the primary transcript to be synthesised ends in a triphosphate grouping. The phosphate at the 5′ end of the primary transcript will in fact be eliminated and there will be fusion with a GMP (coming from a GTP). It is surprising to find that this bond is achieved with phosphoric OH group (anhydride bond) and not a ribose OH. Other modifications may be produced equally, such as a methylation of OH at 2′ of the riboses situated on the two first nucleotides of the 5′ end of the primary transcript.

mRNA will then no longer have a free phosphate 5′ end. It is a fact that there will now be one free OH at 3′ at the two ends.

At present it is thought that this cap, which is in place from the start of transcription, before 30 nucleotides have been assembled, protects the 5′ end of the mRNAs from attack by enzymes (phosphatases, polynucleases). It is equally indispensible to translation, being able to occur in a eukaryote cell. rRNAs and tRNAs are deprived of the cap (and are not translatable). However, the artificial mRNAs, without a cap, can be translated *in vivo*.

(c) *Addition of poly-A to end 3′*

Poly-A is a succession of several adenine nucleotides. This repetitive sequence can exceed 100 nucleotides. It is thought that it:

— assists the passage of the mRNA from the nucleus to the cytoplasm,
— protects the mRNA during translation (the amount of A of poly-A will diminish during translation).

(d) *Maturation of the precursor*

This phase of maturation (called 'processing') takes place in the nucleus. The precursor of mRNA has to undergo many modifications before finally producing mRNA. In particular, a loss of transcribed parts corresponding to the introns occurs. The cutting, or 'excision' of the transcribed introns is followed by a 'splicing', i.e. the remaining segments corresponding to the transcribed exons will be joined or fused end to end.

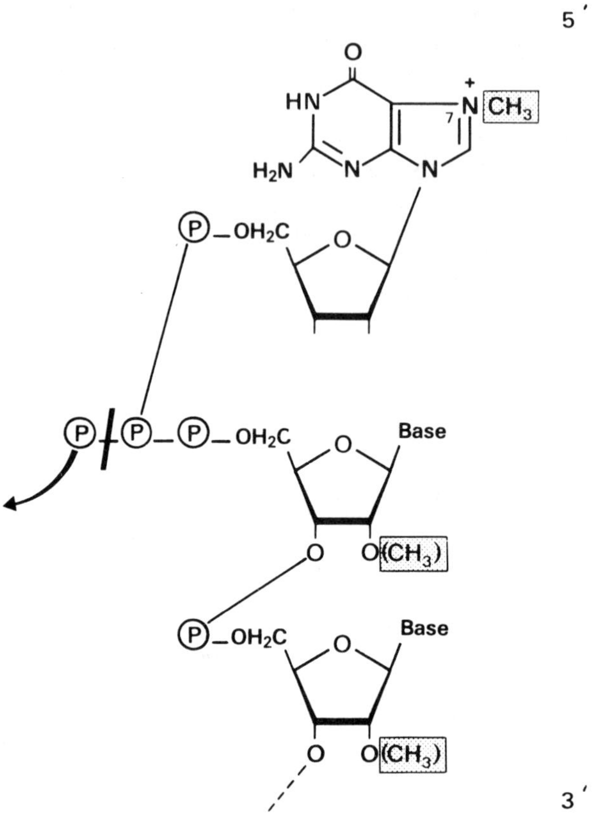

Fig. 54 — The 'cap'.

It is important not to confuse 'introns' (in DNA) with 'transcribed introns' (in the primary transcript). It is better to say that the transcript of introns (rather than the introns) are excised, since one could suppose there to be some rehandling at DNA level. In reality, Gilbert, the inventor of the words introns and exons, defined both the sequence of DNA and the transcribed sequence as 'intron'. The same for the exons — exon coming from 'exist' to indicate that the (transcript of) the exon will emanate from the nucleus and be part of mRNA in the cytoplasm. Some confusion may arise for the non-specialist. This is why the reader is strongly advised to use the terms 'intron' and 'transcript of intron'. The same applies to 'exon' and 'transcript of exon'.

3. How do excisions of intron transcripts and splicing of exon transcripts come about?

(a) Excision–splicing enzymes

- Recognition of the joining of transcribed exon–transcribed intron

Excisions and splicings are achieved by enzymes with a still little-known mechanism. These enzymes, called 'excision–splicing enzymes', do not appear to be specific to

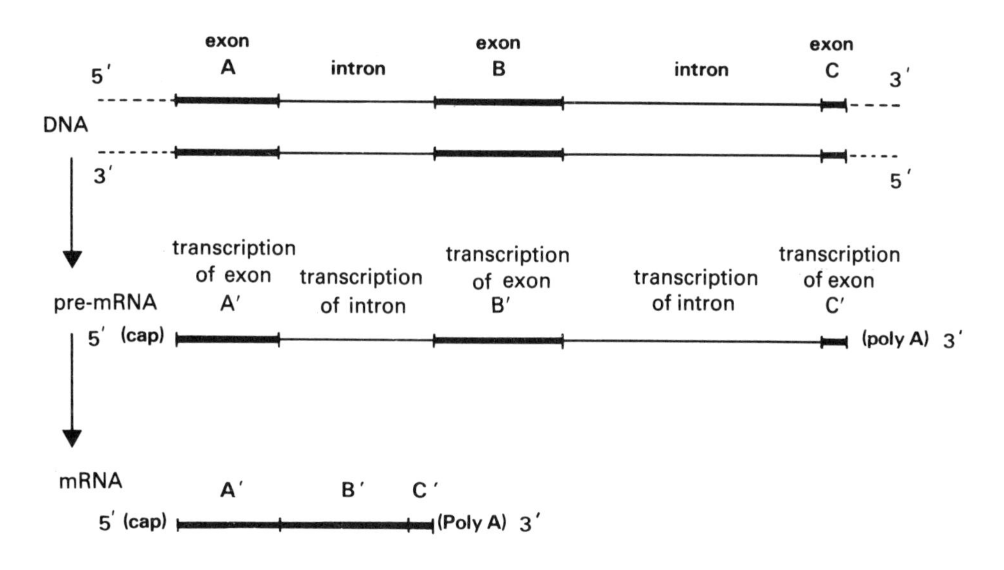

Fig. 55 — Transcription in eukaryotes (e.g. gene having three exons, A, B, C).

the transcript of the intron being investigated. In fact it seems that the enzymes recognise not the intron transcript but the join between that and the exon transcript. It has been noted that the sequences at junction level are very often comparable between one primary transcript and another. In particular the intron transcript most frequently start with the GU sequence and often end with the AG sequence. This excision–splicing phenomenon is very important because one error by a single nucleotide in splicing would alter the reading plan during translation and result in a 'false' protein.

- snRNAs

In the nucleus of eukaryotes there are tiny RNAs (consisting of about a hundred nucleotides) which are called snRNAs (small nuclear RNA) and play a part in excision–splicings. In fact they work in the form of complexes with proteins.

Recent trials carried out *in vitro* (with genes of the globin family) show that the intron transcript would firstly be cut (by excision enzymes) at the junction with the exon (upstream of GU). The snRNAs would be able to play a role in recognising the join. In fact complementary (and antiparallel) bondings are possible between the snRNA and the intron transcript. Thus certain snRNAs possess, for example, close to their 5′ extremity, an AC sequence complementary to GU.

- Liberation of an intron transcript in the shape of a lasso

What happens after the first cleavage between the end of the exon transcript and the start of the intron transcript?

The guanine nucleotide (GMP) situated at the 5′ end of the intron transcript bonds up with an adenine nucleotide (AMP) located about 30 or 40 nucleotides

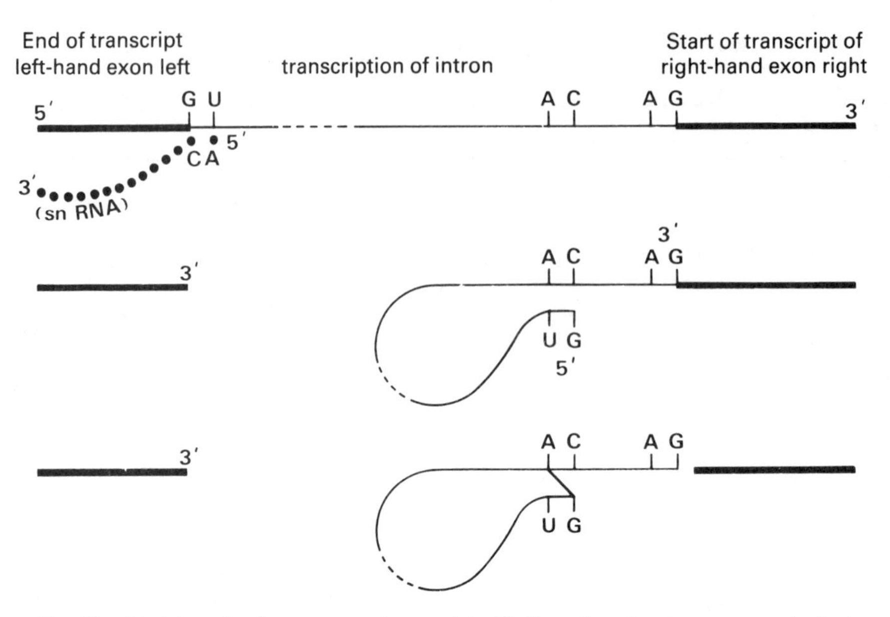

Fig. 56 — Excision of an intron transcript: model with liberation of an intron transcript in the shape of a lasso.

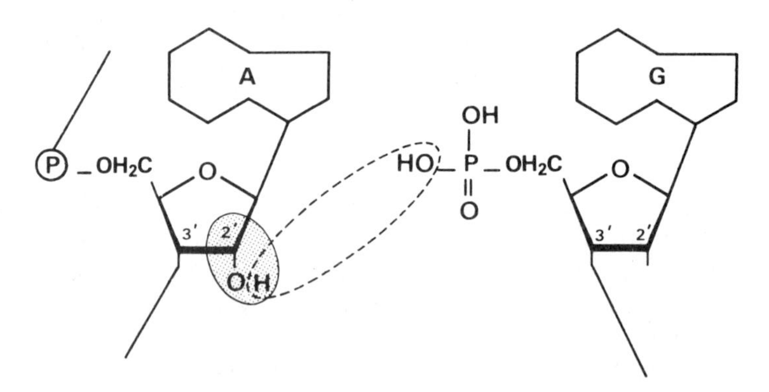

Fig. 57 — Covalent bonds between nucleotides, intervening during the formation of a 'lasso' of the intron transcript.

upstream of the 3′ end of this intron transcript. The phosphoric acid OH of the GMP and the ribose OH at 2′(!) of the AMP implicated in the branching intervene in this bond. At excision of the 3′ extremity, the liberated transcript of the intron is shaped like a 'lasso'. It is probably degraded by nucleases. As for the exon transcripts, they will finally be spliced.

(b) Site of excision–splicing
Elimination of intron transcripts takes place in the nucleus. It is logical that excision of the primary transcript (carrying cap and poly-A) occurs in the nucleus. The primary transcript should be kept apart from the ribosome, as it contains sequences which must not be translated.

(c) Final balance
It is remarkable that only a very small part of the DNA will finally be expressed. The mRNA formed after excision–splicing of the primary transcript corresponds to about 25% of this transcript, which itself corresponds to about 1–5% of DNA. It is evident that only about 1% (or even less than 1%) of DNA will finally code for the proteins!

4. *Role of introns*
Introns have only been discovered in recent years (in the higher eukaryotes). They are not found in eubacteria (*E. coli*) but they have recently been discovered in archeobacteria (which poses some phylogenic problems).

(a) Case of introns coding for a maturase
The recent study of human (and bovine) mitochondrial DNA has enabled us to observe that the DNA of mitochondria is completely devoid of introns! However, it is not the same with yeasts which have introns in their mitochondrial DNA.

Very recently the team led by Slonimski (Gif-sur-Yvette) has demonstrated that a particular intron (intron 2 of the gene of cytochrome b of the yeast mitochondria) gave a protein! This protein in fact originates from the translation not of intron 2 alone, but, more precisely, from that of exon 1+exon 2+ORF of intron 2 (ORF, 'open reading frame'; the portion without a codon stop in the phase with the previous exon). This protein possesses the curious property of excising the transcript of intron 2 (which produced it in the first place). This is why up to now it has not been possible to prove the existence of this 'maturase matricide' which is only present in very small amounts, since it is no sooner formed than it destroys the intron transcript from which it stems — '*traduttore, traditore!*' ('translator, traitor'!), a clear and useful way of saying that the translated protein is a 'traitor', since it destroys the gallant intron transcript which gave it birth).

However, the existence of this maturase can be demonstrated using a mutant strain of yeast where the maturase produced is physiologically inactive. The intron transcript cannot then be excised and it can continue to produce the maturase which accumulates.

This same team has also pinpointed the maturase produced by expression of intron 4. This maturase excises not only its own intron transcript, but also an intron transcript located on the primary transcript emanating from a neighbouring gene (coding for a sub-unit of the cytochrome oxidase).

Although it has thus been shown in a very 'pretty' way, that an intron may be translated into a protein, the role of introns is not always well understood — it is to be hoped that these introns have functions other than merely coding for the proteins excising the transcript of introns!

Moreover it is not yet possible to generalise these results obtained from the mitochondrial genes of yeast.

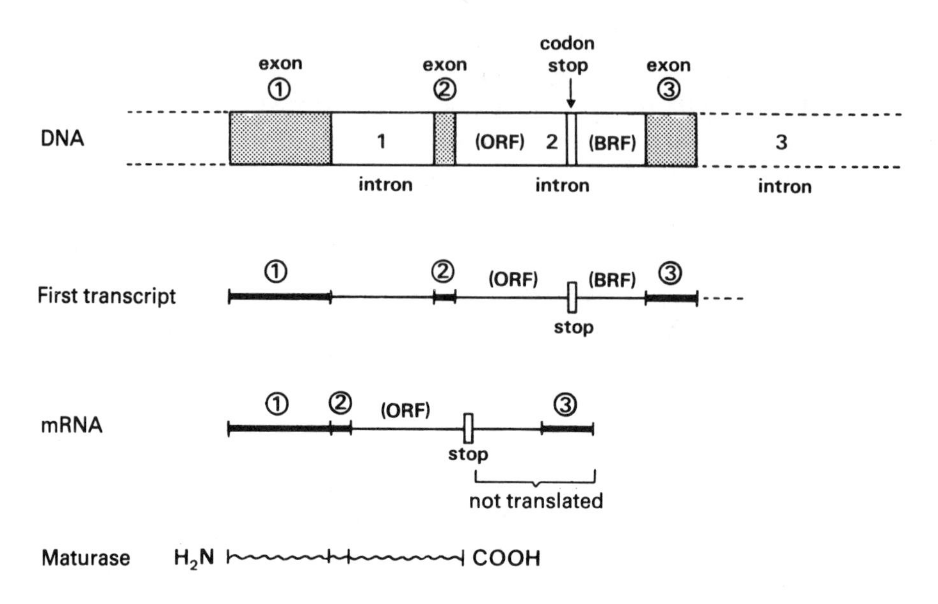

Fig. 58 — Representation of start of the gene coding for cytochrome b in yeast mitochondria (*Saccharomyces cerevisiae,* strain 777-3A). This gene possesses six exons and five introns. Only the first three exons and introns are shown here. The transcript of intron 1 is excised. Translation relating to the transcript of exon 2 (corresponding to the ORF) gives the maturase. This maturase then excises the transcript of intron 2 and facilitates the union of the transcripts of exons 1+2 with the transcript of exon 3. The translation of the exon transcripts will end in the synthesis of cytochrome b. However, the destruction of the transcript of intron 2 by the maturase will arrest the synthesis of new maturase molecules. ORF, 'open reading frame'; BRF, 'blocked reading frame'; ▯, codon stop.

(b) Other hypotheses

A hypothesis for the role of introns has recently been put forward. This hypothesis has been developed as a result of the observation of what are called the 'mosaic proteins'. These are proteins where properties common to other proteins are found, even though these proteins have conversely, a different structure and different functions. (Thus, in the protein called 'the LDL receptor', property 1 responsible for the linking with LDLs shows a great homology with a sequence found in the C9 component of complement. Property 2 likewise shows a homology with a peptide sequence found on four other different proteins). It is thought that in the course of evolution, certain genes were constructed from exons borrowed from other genes, and that the introns, in demarcating the exons, would have facilitated the mobility of the exons in the genome.

The new concept of 'supergene family' regrouping genes (having a common ancestor) inter-related by sequence, has been developed recently by Brown and Goldstein (Nobel Prize for physiology and medicine, 1985).

II TRANSLATION

Translation is the mechanism by which mRNA will be 'decoded'. It is the second phase of peptide synthesis. It succeeds the transcription of DNA and mRNA.

— The terms 'transcription' and 'translation' are easily understood. The word 'transcription' is used because DNA and RNA are copied without the language being changed. With both DNA and RNA we are dealing with a nucleotide language. (Thus, 'transcription' recalls that carried out by monks in the Middle Ages when they recopied sections of the Bible. In the same way, 'to transcribe' in music is to write for a given instrument, a piece of music already composed for another instrument).

The word 'translation' is used because this time the language is changed: with mRNA one 'speaks' nucleotide and now, with the peptide chain formed during translation, one 'speaks' amino acid. For this translation a dictionary is needed, the dictionary being the genetic code.

A. THE GENETIC CODE

1. The three-letter code

The only variable parts of mRNA are the bases, since ribose and phosphoric acid are always the same the whole length of an mRNA sequence. Only the bases therefore are implicated in the genetic code. However, there are only four different bases (A, U, C, and G), and there are 20 different amino acids. How can four bases code for 20 amino acids? Three possibilities can be envisaged:

— First possibility: a one-letter code: $4^1=4$
  U would signify amino acid no. 1.
  C would signify amino acid no. 2, etc. . . .
  Thus this system could only code for four amino acids.
— Second possibility: a two-letter code: $4^2=16$.
  UU would signify amino acid no. 1
  UC would signify amino acid no. 2, etc. . . .
  This system could only code for 16 amino acids. Therefore this possibility is not satisfactory either.
— Third possibility: three-letter code: $4^3=64$.
  UUU would signify amino acid no. 1
  UUC would signify amino acid no. 2, etc. . . .
This system could code for 64 amino acids, which is broadly sufficient.

This excellent initial hypothesis has been effectively confirmed! A three-letter code means that three nucleotides (together called either a 'triplet' or a 'codon') carried on the mRNA will be translated to give an amino acid. For example: AUG is a codon (formed from three nucleotides AMP, UMP and GMP), constituting part of the mRNA and coding for methionine.

Sixty-four codons are therefore used as shown in Table 1. Out of these 64 codons:

— Three codons (UAA, UAG, UGA) are 'non-directional' codons which cannot be translated into amino acids. These codons are in fact signals for the end of the reading and are therefore called 'stop codons'.
— Thus, 61 codons remain (for 20 amino acids). Apart from two cases, Met and Trp, coded by a single codon, the other 18 amino acids are coded by several codons, from two to six (e.g.: the six codons of Leu).

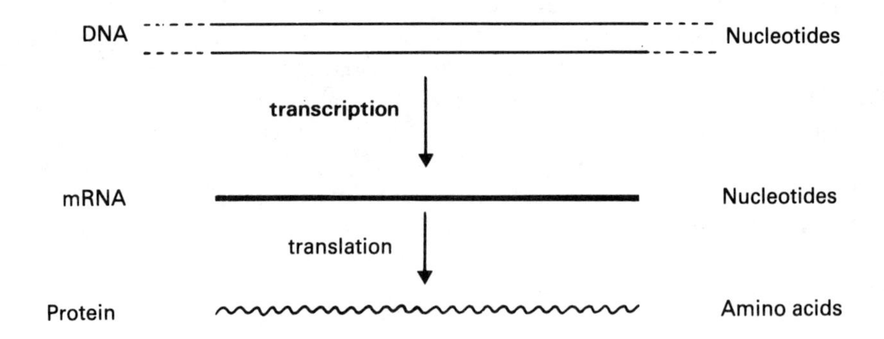

Fig. 59 — The two stages of protein synthesis (transcription and translation).

Table 1 — Genetic code (cytoplasm) codon of the mRNA

First nucleotide	Seceond nucleotide				Third nucleotide
	U	C	A	G	
U	Phe	Ser	Tyr	Cys	U
	Phe	Ser	Tyr	Cys	C
	Leu	Ser	Stop	Stop	A
	Leu	Ser	Stop	Trp	G
C	Leu	Pro	His	Arg	U
	Leu	Pro	His	Arg	C
	Leu	Pro	Gln	Arg	A
	Leu	Pro	Gln	Arg	G
A	Ile	Thr	Asn	Ser	U
	Ile	Thr	Asn	Ser	C
	Ile	Thr	Lys	Arg	A
	Met	Thr	Lys	Arg	G
G	Val	Ala	Asp	Gly	U
	Val	Ala	Asp	Gly	C
	Val	Ala	Glu	Gly	A
	Val	Ala	Glu	Gly	G

2. *Deciphering the genetic code*

The deciphering of the genetic code was an extraordiary discovery made in the 1960s. It is one of the marvels of biochemistry, and may seem even more amazing than succeeding in deciphering hieroglyphics.

To decipher the genetic code a very clever technique was used. It consisted of creating a very special mRNA in the laboratory: the poly-U mRNA consisting of nothing but uracil nucleotides. Placed in contact with ribosomes, amino acids and other factors needed for translation, the poly-U provoked the synthesis of a peptide chain formed uniquely of phenylalanine. This indicated that the codon UUU coded for phenylalanine.

This experiment could be repeated with poly-A, in which case it would be seen that the codon AAA codes for lysine.

Mixed polynucleotides having two different nucleotides repeated, for example poly-AG, could be used.

Thus poly-AG:

AGAGAGAGAGAG will code for (Arg–Glu–Arg–Glu)n.

By itself, however, this result does not enable us to determine if Arg is coded by AGA and Glu by GAG, or vice versa. The trial must be repeated with poly-AAG. This artificial mRNA will give three kinds of polypeptide chains according to whether the decoding begins at AAG (poly-Lys), AGA (poly-Arg), or GAA (poly-Glu). The only codon common to the previous experiment is AGA which thus codes for Arg.

For their success in deciphering the genetic code, Khorana and Nirenberg obtained the Nobel prize for physiology and medicine in 1968.

3. *Characteristics of the code*

(a) *It is universal*

The code is the same in all organisms, be they animal, plant, bacterium or even virus; a fact which is truly extraordinary!

In fact, since 1980, Sanger (twice Nobel prize-winner for chemistry, in 1958 and 1980) and his team have questioned the universality of the genetic code. Their results were obtained with human mitochondrial DNA.

This work, which has been reported in two publications in the journal *Nature*, revealed that the code is no longer as 'universal' as was previously thought: in the DNA of human mitochondria some codons differ!

Thus AUA no longer codes for Ile but for Met
UGA no longer codes for Stop but for Trp
AGA no longer codes for Arg but for Stop, and
AGG no longer codes for Arg but for Stop

Since then, other differences have been found relating to other codons coming not from human mitochondria but from those of yeasts, trypanosomes, etc. Recently an exception has even been found which is not in mitochondria. The codon UAA of cytoplasmic mRNA of paramecia (protozoa) is not a stop codon but a code for Gln! However, it remains no less true that the code is universal (albeit with some very rare exceptions).

(b) It is degenerate

The code is said to be 'degenerate' because the same amino acid can be coded by several different codons. It should be noted that in numerous cases the triplets coding for the same amino acid do not differ among themselves except at the third base. Take for example the codons of serine: UCU, UCC, UCA, UCG.

The term 'degenerate' has a pejorative connotation, although on the contrary, there are cases where it can be a definite advantage. In effect, the degeneracy constitutes a system of protection against mutations which might be produced. For example, in the case of serine codons, a mutation on the third base would have no effect. We shall see, however, in the chapter on genetic engineering, that the degeneracy of the code represents an enormous inconvenience for the biochemist who is synthesising molecular probes.

(c) It is (non)-overlapping

Until the last few years, the code was said to be non-overlapping, that is to say that the part of a DNA strand which was expressed, was transcribed and then translated regularly triplet by triplet, in one single reading frame. It has been said for some years, however, that there are known cases where the transcripts of the same region of DNA may be read according to several reading frames (after reorientation by a base). For example, in theory the following sequence:

UUUACGAUGUA

can be translated according to a first frame:

UUU ACG AUG UA
Phe–Thr–Met

or after the offsetting by a base, according to a second frame:

UUU CGA UGU A
Leu–Arg–Cys

or after a further offsetting by a base, according to a third frame:

UAC GAU GUA
Tyr–Asp–Val

A further shifting would bring the translation back to the initial frame.

The first case described was that of the virus ϕX174. It was very surprising to discover that, of the nine genes of this virus, two are small genes completely covered by larger ones (gene B covered by A and gene E covered by D). The overlaid genes were read according to two different reading frames. Here, an identical DNA sequence carries the information needed for the synthesis of two proteins having totally different acid sequences! An extraordinary example of genetic economy!

Amazement does not end there. Take the example of genes D and E. It is noted that the codons corresponding to gene D often contain U at the third position. Given that, for the codons corresponding to gene E, the reading frame is shifted by one base to the right, they will therefore have U at the second position numerous times. Now, if one examines the genetic code, it can be seen that codons having U at the second position correspond to hydrophobic amino acids. The gene E then codes for a protein which would have properties close to those of a detergent. Effectively, the role of this viral protein is to break down the wall of the host cell (thus provoking the lysis of the bacteria where the viruses multiply).

Since then, other overlappings of genes have been discovered. For example, there is the case of the hepatitis B virus, where a long sequence of DNA called 'region P' partially overlaps three other sequences (that coding for HB antigens, that coding for the HBc antigen and that coding for peptide X).

B. Site of translation

Translation takes place in the cytoplasm, at ribosome level. The ribosome is the 'reception area' and receives the elements needed for protein synthesis. These are:

— the tRNAs which supply the amino acids,
— the mRNA.

In addition it contains, in the large sub-unit, the 'machine tool'; the enzyme which will successively 'hook' an extra amino acid onto the peptide chain in the course of its elaboration. This enzyme activity is called 'peptide transferase'.

C. The necessary elements

1. Amino acids

Amino acids are the 'bricks', the 'prime material' of a peptide chain. The peptide chain is in fact made up of a succession of amino acids linked one to another by amide bonds, formed by the elimination of a molecule of water between the NH_2 of an amino acid, and the COOH of another amino acid.

In fact, it is not unimportant which –COOH, or which $-NH_2$, that might happen to be present in these amino acids, takes part in these bonds. Each time it is –COOH carried by the α-carbon of an amino acid, and the $-NH_2$ carried by the α-carbon of another amino acid. It is this amide bond, particularly between two amino acids, which is called the peptide bond.

2. mRNA

The bonding of amino acids into a peptide chain should take place in a certain order. This occurs thanks to an mRNA. It is in fact the latter which, as we have seen, carries the DNA message: it indicates the order in which the different amino acids will succeed each other.

However, the amino acids do not assemble directly on the mRNA; an adaptor is needed.

3. tRNA

tRNA may be considered as an adaptor between mRNA and the amino acid. It is in fact linked:

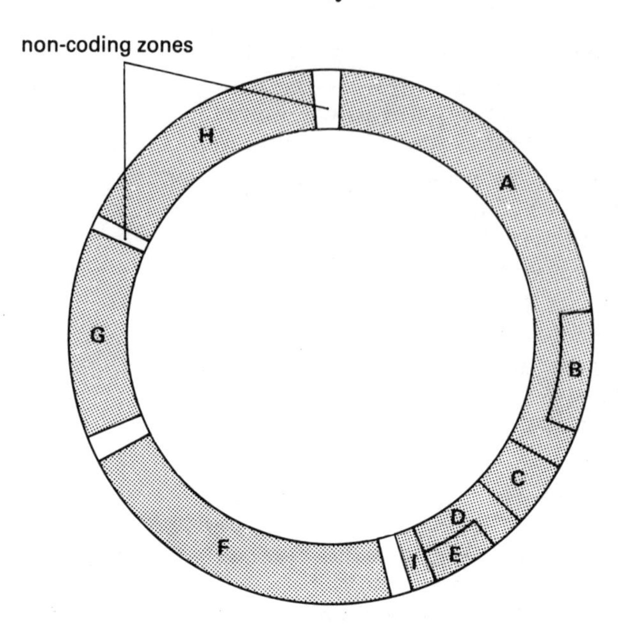

Fig. 60 — Genome of ØX174 (From Fiddes, J. C. 'The nucleotide sequence of a viral DNA' *Science,* Feb 1978, p. 120, Fig. 14.)

— on one side (by its 3′ end: CCA) to the amino acid. This is a covalent ester bond, ('rich in energy') between the tRNA and the amino acid.
— on the other side (by its anticodon) to mRNA. These are weak hydrogen bonds between the complementary bases of the anticodon and the codon.

The amino acids, on standby in the cytoplasm, will not be free when they arrive in the ribosome, but bonded to the tRNA which carries them. We have already seen how these aminoacyl~tRNAs (aa~tRNA) are formed.

D. The different phases of translation

1. Initiation

Close to the phosphate 5′ end of the mRNA is a 'signal codon' which indicates that translation must begin. It is called 'the initiator codon'. This codon is always AUG. It codes for methionine. All the peptide chains therefore, after their synthesis, have methionine as their first amino acid. (As we shall see later, this methionine will usually be removed just after synthesis of the peptide chain.)

Two different tRNAs exist (but having the same anticodon) to transport Met, whether it be an initial Met or a Met destined to be incorporated into the chain. In *E. coli* there are also two different tRNAs to transport an initial Met and an internal Met, but here the initial Met does not have a free NH_2 grouping. The NH_2 is blocked by formic acid which gives formylmethionine (fMet).

We shall try to analyse the mechanism of translation in several stages illustrated by 11 diagrams (Fig. 64).

Just before translation begins, the ribosome is not constituted. The two ribosomic

GENETIC CODE

First nucleotide	Second nucleotide: U	C	A	G	Third nucleotide
U	Phe	Ser	Tyr	Cys	U
	,,	,,	,,	,,	C
	Leu	,,	Stop	Stop	A
	,,	,,	Stop	Trp	G
C	,,	Pro	His	Arg	U
	,,	,,	,,	,,	C
	,,	,,	Gln	,,	A
	,,	,,	,,	,,	G
A	Ile	Thr	Asn	Ser	U
	,,	,,	,,	,,	C
	,,	,,	Lys	Arg	A
	Met	,,	,,	,,	G
G	Val	Ala	Asp	Gly	U
	,,	,,	,,	,,	C
	,,	,,	Glu	,,	A
	,,	,,	,,	,,	G

Fig. 61 — The codon having U in second position coding for hydrophobic amino acids.

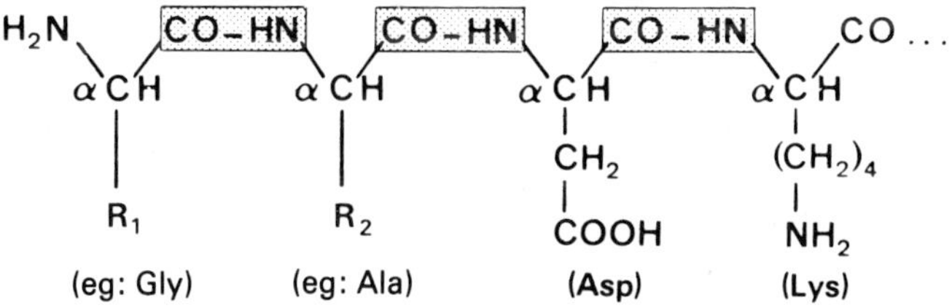

H_2N — αCH (R_1, eg: Gly) — CO–HN — αCH (R_2, eg: Ala) — CO–HN — αCH (CH_2–COOH, Asp) — CO–HN — αCH ($(CH_2)_4$–NH_2, Lys) — CO ...

Fig. 62 — Peptide bond (between two amino acids).

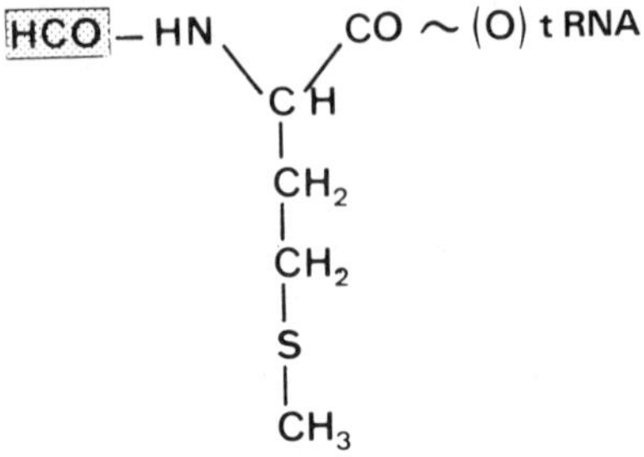

HCO – HN CO ∼ (O) t RNA
CH
|
CH_2
|
CH_2
|
S
|
CH_3

Fig. 63 — Formylmethionine (transported by its tRNA).

sub-units are in fact dissociated and free in the cytoplasm. At the initial phase the small sub-unit forms a complex with:

— mRNA at codon AUG (initiator codon), and
— tRNA carrying the initial methionine.

The large sub-unit is then added. The ribosome is now constituted and functional. Ribosomes have two sites to combine the tRNAs:

— site A (amino acid site) to which the tRNA (the carrier of the amino acid) will come
— site P (peptide site) for the tRNA carrier of the peptide chain in the course of elongation.

The tRNA which transports the initial methionine has a conformation which allows it to lodge at the peptide site.

2. Elongation

After initiation, the first amino acid is in place. A peptide bond will now have to be formed during the following phase, known as elongation.

For each amino acid to be hooked on, i.e. for each peptide bond to be manufactured, an identical, three-phase cycle is performed each time.

(a) First phase: hooking a new aminoacyl∼tRNA (aa∼tRNA) onto the ribosome

The second tRNA arrives at site A of the large sub-unit with amino acid no. 2 (amino acid site). It is codon no. 2, located on the mRNA after codon AUG, which then determines the second anticodon, i.e. the second tRNA, and thus the second amino acid.

(b) Second phase: formation of the peptide bond

There is a rupture of the bond between the methionine and the first tRNA (which is ejected). Thus, the peptide bond is formed between:

— COOH of amino acid no. 1 (Met) and

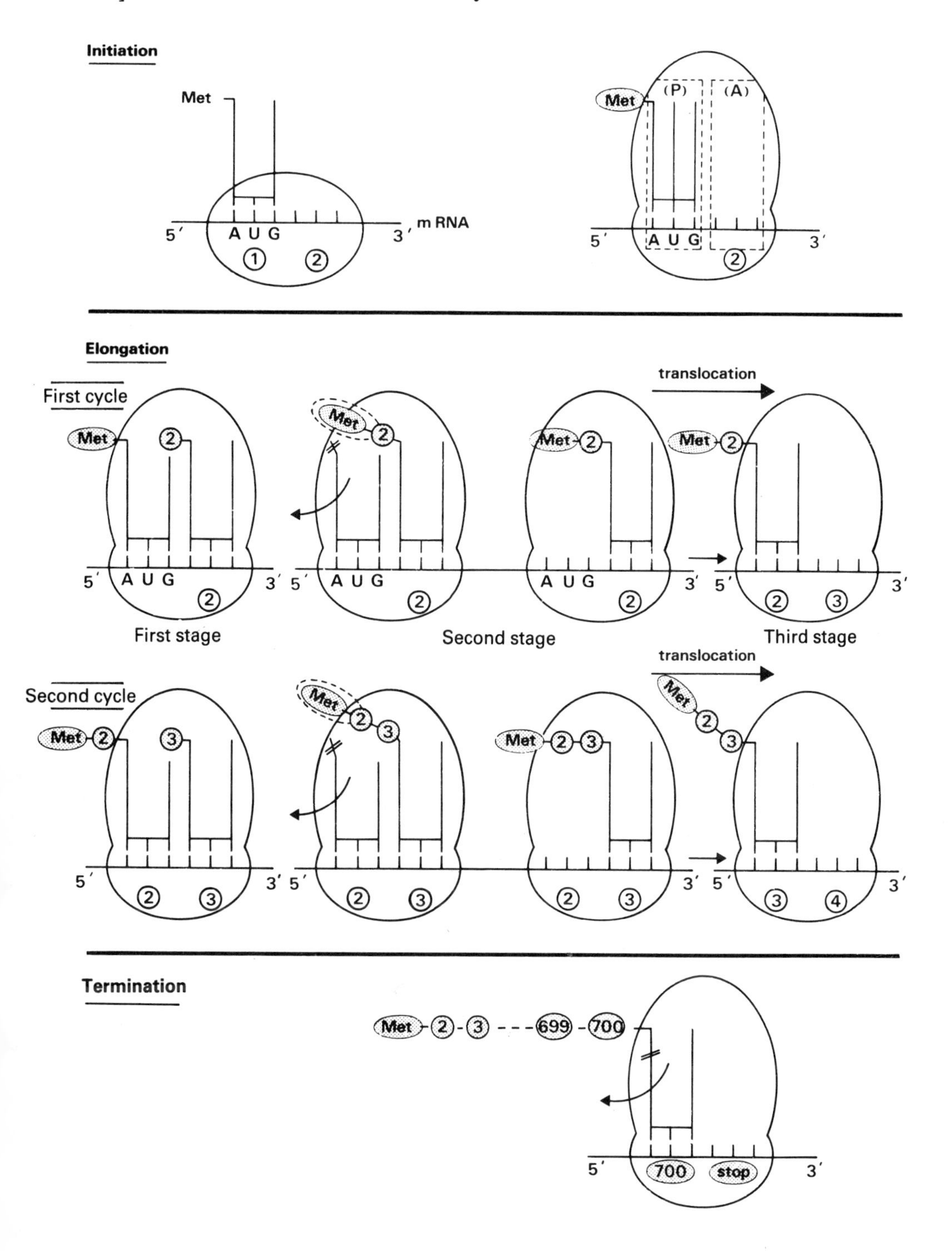

Fig. 64 — The different stages of translation.

— NH_2 of amino acid no. 2 carried by rRNA no. 2.

In fact, as we shall see, the –COOH of the first amino acid was not free, since it was already bonded to tRNA.

The formation of the peptide bond to yield the dipeptide (now hooked onto the second tRNA) and the detaching of the first tRNA occur simultaneously. Remember that the enzyme which intervenes in this bonding is called the peptidyl transferase.

At this stage, a dipeptide has been formed which is lodged at the amino acid site, where it is carried by tRNA no. 2. The ejection of the first tRNA has vacated the peptide site.

(c) Third phase: translocation

The ribosome will advance by a notch on the mRNA in the direction $5'\rightarrow3'$. A notch means three nucleotides (or rather, one codon). A new codon (codon no. 3) is now in place opposite the amino acid site.

Simultaneously, tRNA no. 2 which carried the dipeptide (formed by methionine and amino acid no. 2) has passed from site A (amino acid site) to site P (peptide site). From this change of site comes the name 'translocation'. However, it is still opposite its codon (codon no. 2) since the ribosome has advanced by a notch.

Numerous cycles will follow one another each time with the three phases:

— hooking of aa~tRNA onto the ribosome
— formation of the peptide linkage
— translocation.

Take the example of a protein having a molecular weight of 70 000 and made up of about 700 amino acids (an amino acid has a molecular weight of approximately 100). To form this protein, 699 amino acids will have to attach themselves successively, i.e. 699 cycles, will be performed. Each time round the cycle, the peptidyl transferase, as its name indicates, will transfer the peptide linked to the former tRNA (in site P) to the new tRNA (in site A). It cuts, transfers and fuses. Each time, therefore, the peptide increases by the full length of an amino acid.

The ribosome may be compared to a decoder or reading head which is displaced on the mRNA always in the $5'\rightarrow3'$ direction, three nucleotides by three nucleotides, codon by codon.

3. Termination

The end of translocation occurs when the ribosome, in advancing by a notch on the mRNA, finds a stop codon. There are three stop codons, UAA, UAG, UGA. These codons do not code for any amino acid. No tRNA exists which has a complementary anticodon to any of these three codons. Thus, no tRNA will come to site A. A separation of the last tRNA from the peptide chain will therefore occur. The ester bond uniting this last tRNA to the last amino acid of the peptide chain is hydrolysed by peptidyl transferase, thus liberating the peptide chain.

The ribosome breaks down again into two sub-units which can begin new readings of mRNA.

E. Energy balance

1. For the initiation phase

For the formation of the small sub-unit complex — mRNA–aa~tRNA — the following are involved:

— one molecule of GTP, which will be hydrolysed to GDP,
— proteins which are called 'initiation factors'.

For simplicity however, we shall not consider the balance of this phase further. Initiation only occurs once during synthesis of a peptide chain, in order to correctly position the tRNA carrying the initial methionine — in the site (peptide) of the ribosome.

2. For the elongation phase

Let us turn our attention to elongation, the phase which recurs each time a new amino acid becomes attached, and thus several hundred times for one protein (for example one having a molecular weight of 70 000 and thus containing about 700 amino acids).

What does each amino acid attachment cost? The energy needed to form a peptide bond is equivalent to at least four bonds rich in energy. The peptide bonds are formed using the energy-rich bonds contained in ATP and GTP.

(a) ATP, GTP and elongation factors

• Hydrolysis of one ATP

In order to increase one peptide chain by one unit, i.e. by one amino acid, we have seen that one amino acid is not attached directly to the last amino acid of this peptide chain, but in fact an aa~tRNA is attached to the last amino acid of the peptidyl~tRNA in the course of formation. In order to attach one amino acid it would be necessary to have one aa~tRNA to supply both an amino acid and the energy required.

However, to obtain one aa~tRNA we have seen that initially one aa~AMP is formed (the amino acid then being transferred from the AMP to the tRNA.) Remember that the formation of one aminoacyl~AMP consumes two energy-rich linkages.

One molecule of ATP contains two energy-rich linkages, and can, in this example, be hydrolysed to ADP+P_i or to AMP+PP_i. The hydrolysis of ATP to ADP and P_i (as for example is the case with the phosphorylation of glucose to glucose-6-phosphate by hexokinase) would not allow an adequate energy supply. On the contrary, the hydrolysis of ATP to AMP+PP_i, followed by the hydrolysis of the pyrophosphate liberated by a pyrophosphatase, supplies the energy needed for the reaction, and moreover renders it irreversible.

Thus it is already possible to confirm that the formation of the aminoacyl~tRNA costs two energy-rich linkages.

$$aa + tRNA + ATP \rightarrow aa \sim tRNA + AMP + 2\,Pi$$

In fact it may be remarked that it is not the energy-rich linkage between an amino acid (for example no. 5) and its tRNA which will be utilised to unite this amino acid (no. 5) to the final amino acid (no. 4) of the peptide chain. The linkage energy (of amino acid no. 5) will in reality be conserved, and used in the following phase to unite the next amino acid (no. 6).

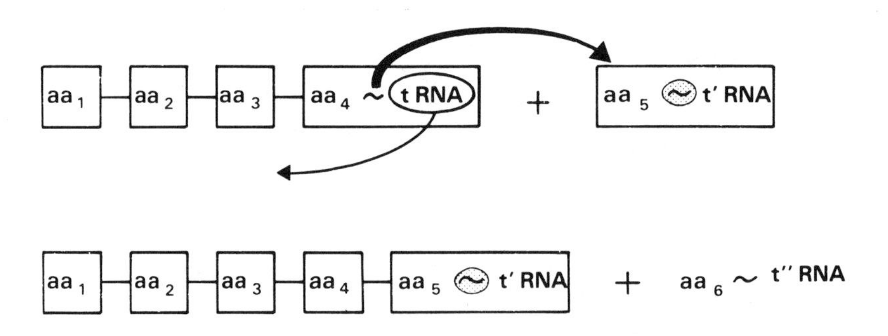

Fig. 65 — Origin of the energy needed to form the peptide bond.

Thus, at each cycle, the energy needed to form the peptide bond comes not from the ester linkage of aminoacyl~tRNA situated in site A, but from the hydrolysis of the peptidyl~tRNA located in site P.

This comment does not change the energy balance at all: to form a peptide bond, one energy-rich bond coming from an aa~tRNA is required. However, to form this aa~tRNA it has in fact been necessary to consume two energy-rich bonds originating from the ATP.

We shall see that two other energy-rich bonds are still needed.

- Hydrolysis of two GTPs

In the course of the attachment of each amino acid, i.e. during each elongation cycle, two GTP molecules will be involved. Each GTP will be hydrolysed as follows:

$$GTP \rightarrow GDP + P_i$$

It is thought that these two GTP molecules act as allosteric factors.

- Elongation factors (family of G proteins)

Proteins called 'elongation factors' play an important role. These proteins belong to the family of 'G proteins' and will be mentioned again in the chapter on oncogenes.

These are proteins which have the property of only being active when they are linked to GTP. The action made possible by the G protein–GTP bond varies for each protein of this family. However, this action will always be very brief since all these proteins also have a GTPasic activity. Thus, as soon as a G protein unites with a GTP, an action specific to this protein is immediately triggered but the GTP responsible for the action of this protein will also be hydrolysed. GTP behaves as an allosteric factor. When it is linked to the G protein it modifies the conformation of the protein which then becomes active. When GTP is hydrolysed, protein G remains bonded to GDP; however, GDP does not have the same properties as GTP and cannot maintain protein G in an active conformation. There is then a return to the initial conformation of protein G. Thus it is clear that a G protein, in combining with GTP, acquires a transient function. Let us now return to the elongation factors and the two hydrolysed GTP molecules.

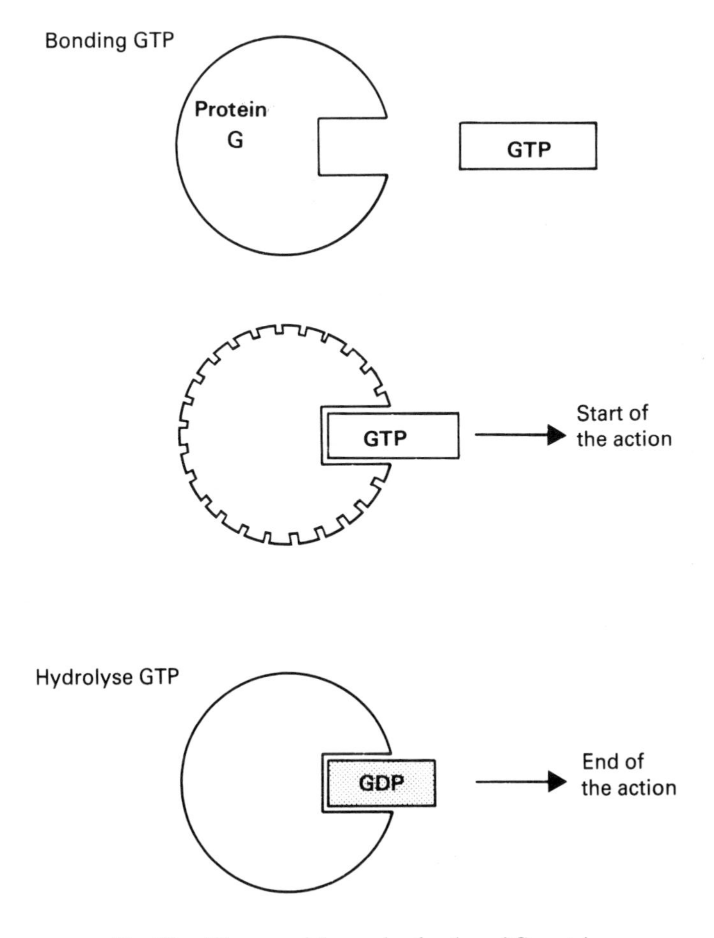

Fig. 66 — Diagram of the mode of action of G proteins.

(b) Intervention of GTP and elongation factors

- *During positioning of the amino acid in site A of the ribosome*

The first molecule of GTP and the first elongation factor are used at the moment of amino acid positioning in site A of the ribosome.

GTP provokes a change in the spatial structure of the eleongation factor thus allowing the elongation factor to attach itself to the ribosome and perform its role, i.e. to fix the aa~tRNA in site A of the ribosome. The elongation factor (EF-Tu in prokaryotes, EF-1 in eukaryotes) combines aa~tRNA and the ribosome only if it is bound to GTP.

GTP is then hydrolysed to GDP+P_i. As we have seen, GDP has not the same ability as GTP in maintaining the conformation of the elongation factor. The hydrolysis of GTP to GDP provokes the return of the elongation factor to its initial conformation. In its initial conformation the elongation factor can no longer remain on the ribosome and becomes detached. It is likewise detached from the aa~tRNA, which is now fixed in the ribosome.

Thus, in order to position the 'good' tRNA correctly in site A of the ribosome, one GTP has had to be hydrolysed to GDP. Balance: one energy-rich linkage consumed.

- During translocation

A second GTP molecule is used at the moment of translocation. In order that the translocation may be effected, an elongation factor is also required (different from the preceding one). Currently it is thought that it is the attachment of this factor to the ribosome which provokes translocation.

As in the preceding case, in order that the elongation factor may become fixed to the ribosome, a molecule of GTP is required. It is thought that, here too, GTP provokes a modification of the spatial structure of this second elongation factor, thus allowing it to hook onto the ribosome. In the same way, after the elongation factor has acted (and has thus provoked translocation), the GTP is hydrolysed to GDP. Again the hydrolysis of GTP provokes the return to the initial conformation of the elongation factor which then becomes detached from the ribosome.

Translocation, therefore utilises an energy-rich linkage originating from one GTP.

3. For the termination phase

Intervening at this final phase will be:

— one GTP,
— proteins called 'termination factors'.

Given that termination (like initiation) only occurs once in the course of synthesis of a peptide chain, we shall ignore the energy balances of this phase.

Summary

For each amide linkage it is necessary to provide at least four energy-rich linkages (two coming from a molecule of ATP and two from two molecules of GTP). Now, to make an amide bond, a single rich bond would theoretically suffice. Four energy-rich bonds is too much to pay! Nevertheless, that it is the price which the cell has to pay in order to guarantee the accuracy of the translation: to read the message easily, to position the amino acid correctly, to jump a notch. To make a peptide bond is not simply to make an amide bond, it is equally to respect a sequence of amino acids.

F. Comments on

1. Polysomes

An mRNA can be read by several ribosomes at once. The ribosomes hooked onto an identical mRNA molecule form what is called a 'polysome' (or polyribosome). This increases the output, since for each ribosome travelling along an mRNA there will be a synthesised peptide chain. The distance which separates two ribosomes is about 100 nucleotides. The closer a ribosome is to the 5′ end, the shorter will be the chain being formed. Conversely, the nearer the ribosome to the 3′ end, the longer the chain in the course of formation.

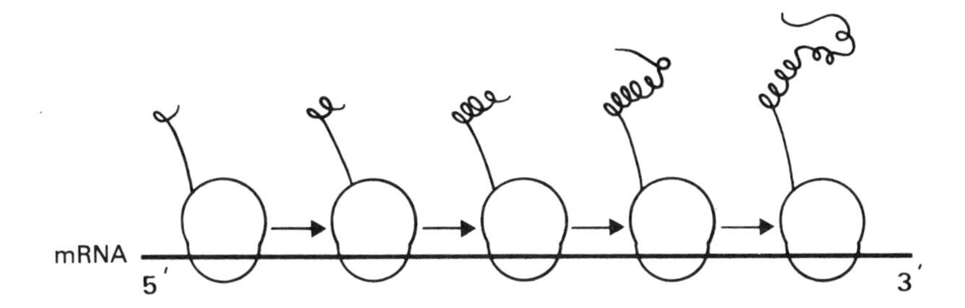

Fig. 67 — A polysome.

The fact that an mRNA strand may be read by several ribosomes allows numerous peptide chains to be formed from a single mRNA.

2. The 'cap'

In prokaryotes (where mRNA does not have a cap) the ribosome brings about the translation of the mRNA from AUG as far as the first stop codon. However, if there is an AUG codon nearby it can follow the translation and give another protein different to the first. In this case it is said that mRNA is polycistronic.

In eukaryotes (where the mRNA possesses a cap needed to effect the translation) translation will begin at the first AUG codon (in phase) and will end at the first stop codon encountered (in phase). The ribosome will be unable to resume its reading of the mRNA beyond the first stop codon, even if an AUG codon is present close by. The mRNA of eukaryotes is said to be monocistronic.

Translation in prokaryotes:

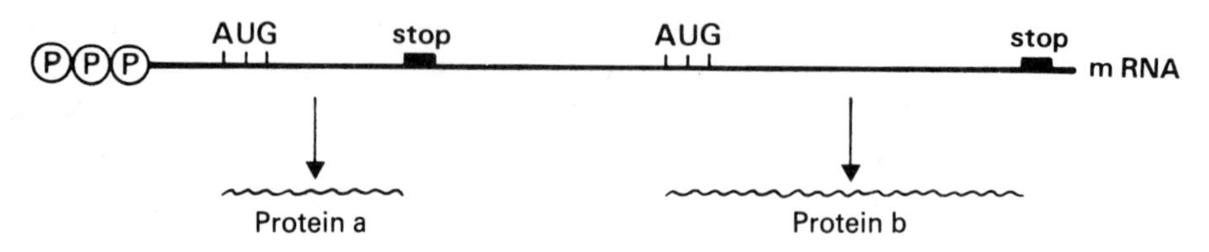

Translation in eukaryotes:

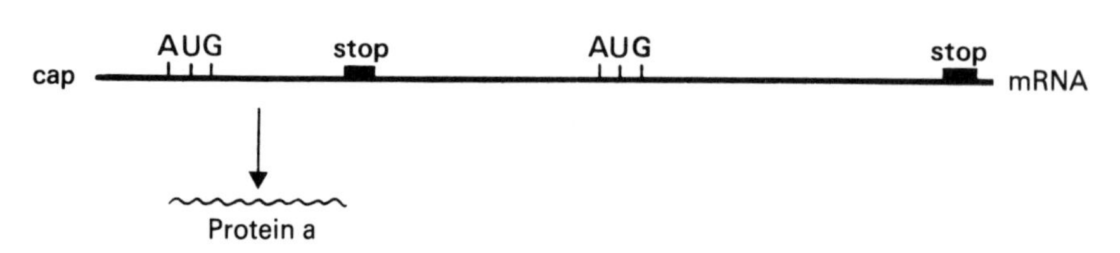

Fig. 68 — Polycistronic or monocistronic mRNA.

G. The 'wobble' or 'fluctuating base'

1. What do we call base 'wobble'?

We have seen that the genetic code contains 64 codons, three of which are stop codons, so that there are 61 codons coding for 20 amino acids.

It could be expected then that there are 61 tRNAs (because of the rule of codon–anticodon complementarity). In fact, according to Crick's hypothesis of wobble, 32 tRNAs could be sufficient to transport the 20 different amino acids. This is explained by the fact that a tRNA can recognise several different codons coding for the same amino acid (N.B. a given tRNA will never recognise the codons for two different amino acids).

The 'wobble' base (or fluctuating base) is the first base of the anticodon (on the tRNA), facing the third base of the codon (carried by the mRNA).

The first two nucleotides of the mRNA are strictly complementary to the anticodon, in accordance with the classic rule of pairing.

The same does not apply to the third mRNA base. In effect, the pairing of this base with the first base of the anticodon can occur in a non-classic fashion. This is what we shall see appearing under the name of: 'pairing wobble'.

2. The different types of wobble: I/UCA, U/G and G/U (and U/N in mitochondria)

(a) In cytoplasm

There are three kinds of pairing wobble: I/UCA, U/G and G/U.

- I/UCA wobble

I is the abbreviation used for IMP (hypoxanthine nucleotide). When I is situated in a wobble position on the anticodon, I can pair equally well with U or C or A (situated in the third position on the mRNA codon). The links between the other two base pairs take place in the classic manner.

Let us take as an example the case of glycine. Gly is coded by four different codons: GGU, GGC, GGA and GGG. Consider the first three codons, as well as the three corresponding anticodons: with I in the wobble position and CC in the second and third positions, since these are the complementary nucleotides of GG, it is not necessary to use three tRNAs, to recognise these three different codons of Gly. A single tRNA having 3′-CCI-5′ as anticodon is sufficient to transport Gly and to recognise the three different codons. (It will, however, be necessary to have a second tRNA to recognise the fourth GGG codon).

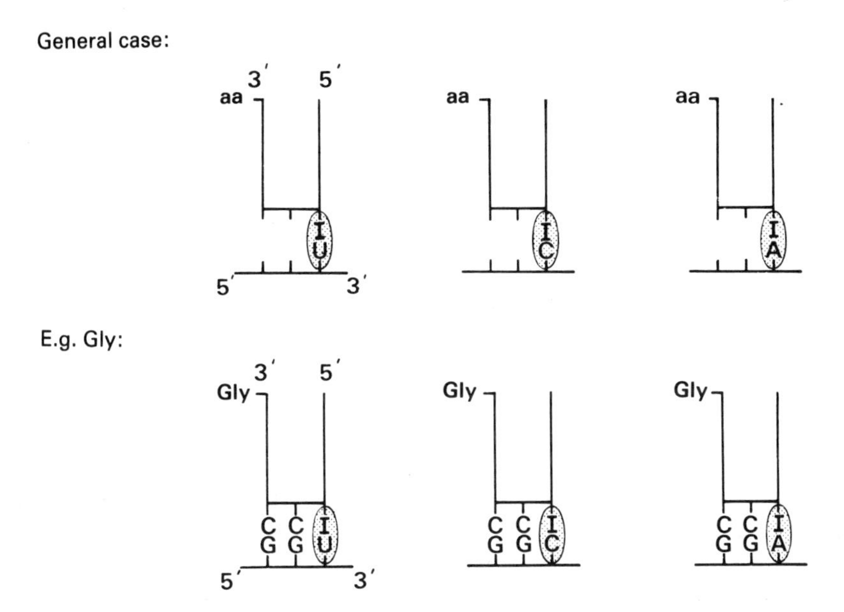

Fig. 69 — I/UCA wobble.

- U/G wobble and G/U wobble

— U/G wobble

When U occupies the wobble position on the anticodon, U can be paired equally well with A (classic pairing) as with G (wobble pairing). As in the previous case, the pairings of the other two base-pairs takes place in the classic way.

Let us take arginine as an example. Arg is coded for by six codons. Let us for the moment look at codons AGA and AGG. Consider these two codons, as well as the corresponding anticodons: with U in the wobble position and with C and U in second and third position, since these are the complementary nucleotides of G and A, a single tRNA having UCU as anticodon suffices to recognise two different codons (AGA and AGG).

— G/U wobble

In a way comparable to the previous case, when G occupies the wobble position on the anticodon, G can be paired equally well with C (classic pairing) or U (wobble pairing).

General case:

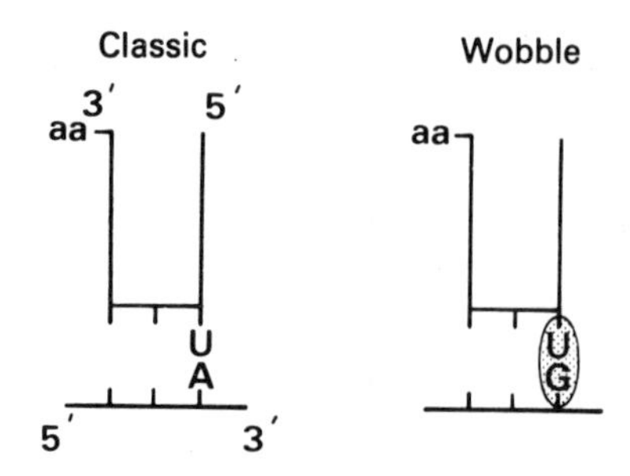

E.g. Arg:

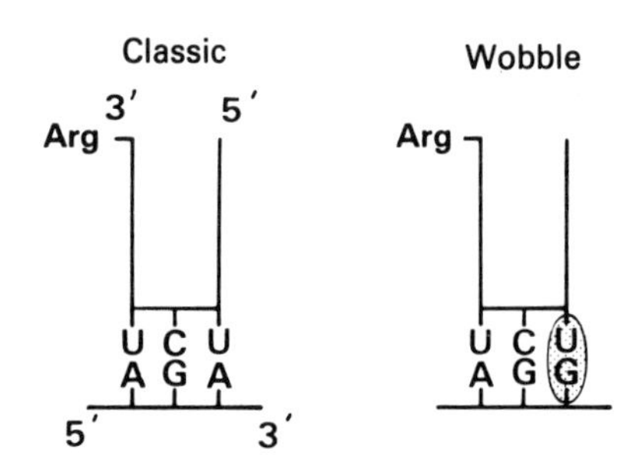

Fig. 70 — U/G wobble.

Let us take asparagine as an example, which is coded for by the codons AAC and AAU, and consider these codons as well as the corresponding anticodons with G in the wobble position. The same 3′ UUG-5′ anticodon (on the same tRNA) recognises AAC (classic pairing) and AAU (wobble pairing) equally well.

In an earlier chapter we saw that guanine in the wobble position can, for certain tRNAs, be replaced by queuine. Queuine pairs with C and even better with U.

The first base of an anticodon (wobble base) therefore determines if one tRNA recognises one, two or three codons. (Quite clearly the three codons considered at Fig. 72(a) are the same bases in the first and second position and these will thus not be represented in this diagram. The same applies to the two codons at (b) and (c).

Thus, I recognises three different codons (at third base),

U and G recognise two different codons,

C only recognises one codon.

- Due to the wobble, 32 tRNAs could suffice

Let us return to the genetic code and apply the name ‘box of four’ to all cases where there are four codons coding for the same amino acid. (e.g. the four codons of Val).

Let us give the name ‘box of three’ to the upper portion of a box of four when three codons code for the same amino acid (while the fourth codon codes for a different amino acid). There is only one case of this, that of Ile.

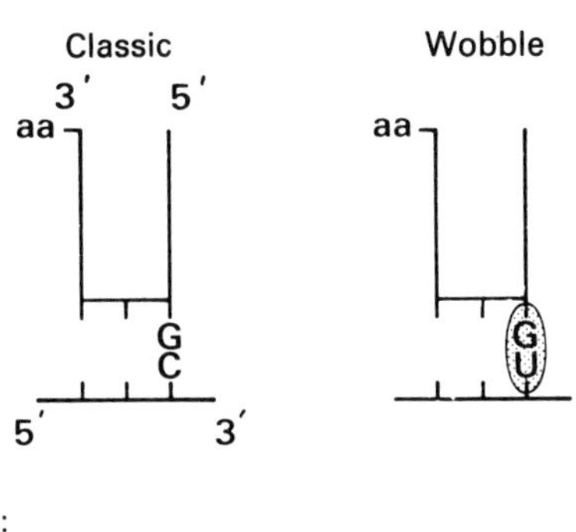

— E.g. Asn:

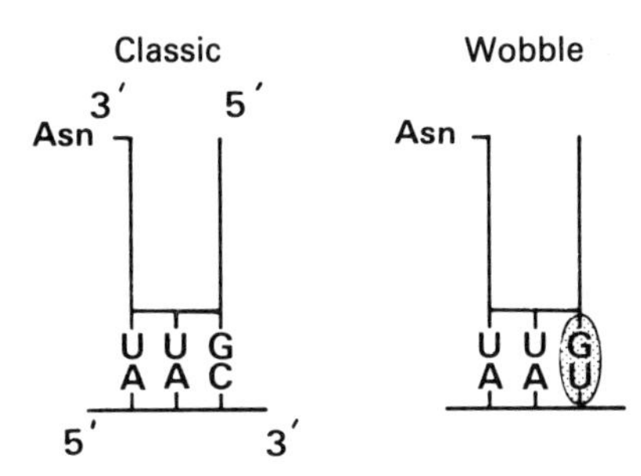

Fig. 71 — G/U wobble.

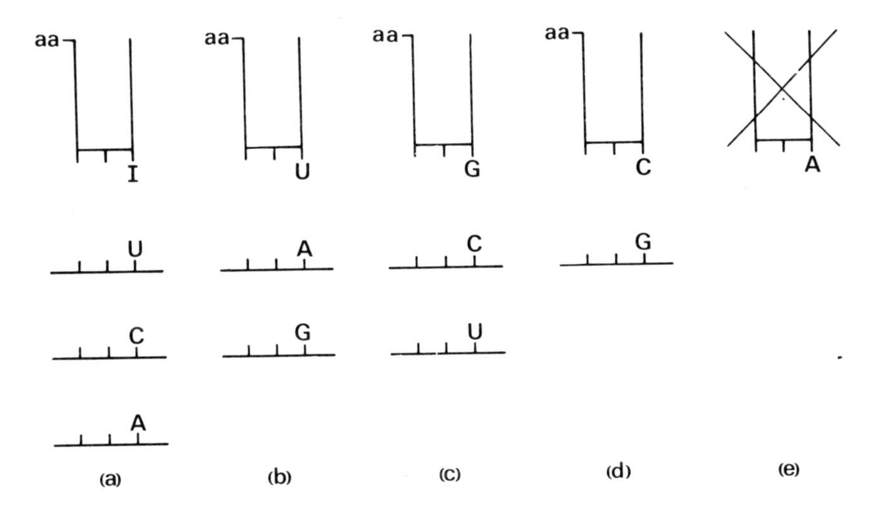

Fig. 72 — Number of codons recognised according to different types of wobble.

Lastly we have an 'upper box of two' when two codons code for the same amino acid and are situated high up in a box (e.g. Phe) and a 'lower box of two' when two codons code for the same amino acid and are situated low down in the box (e.g. Gln).

Now that we know the 'rules of the game' let us count the minimum number of tRNAs for recognising 61 codons.

Beginning with Phe's upper box of two and going along the genetic code from top to bottom, Phe is coded for by codons UUU and UUC. Which is the wobble base capable of recognising U and C at the same time? Let us refer back to the previous diagram (Fig. 72) Answer: G; a single tRNA having G in the wobble position is sufficient to recognise these two codons (the anticodon of this tRNA will thus be 3′-AAG-5′).

Let us continue with Leu (lower box of two) coded for by UUA and UUG. Which wobble base will recognise A and G? It is U, so only a single tRNA having U in the wobble position is needed to recognise these two codons (the anticodon being 3′ AAU 5′).

Let us change boxes. There are still four codons for Leu (=box of four). There will have to be two tRNAs. One will have I in the wobble position to recognise the three codons — CUU, CUC and CUA — situated in the upper part of the box. The other will have C in the wobble position to recognise the codon CUG situated in the lower part of the box.

It is therefore seen that each time there will be need for:

1st One tRNA for each upper box of two.
So for Phe, Tyr, His, Asn, Asp, Cys, and Ser, there should be a total of seven tRNAs.

2nd One tRNA for each lower box of two.
So for Leu (UUA and UUG), Gln, Lys, Glu and Arg (AGA) and AGG), there should be a total of five tRNAs.

3rd Two tRNAs for each box of four.
A first tRNA will have I in the wobble position and will recognise the three upper codons.

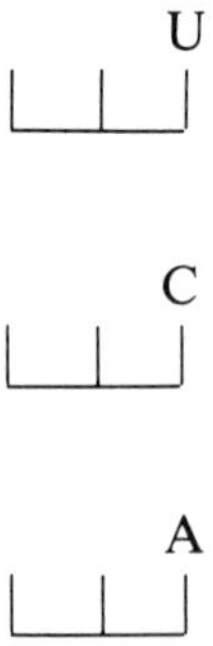

U

C

A

A second tRNA with C in the wobble position will be needed to recognise the lower codon.

G

|_|_|

So for Leu, Val, Ser, Pro, Thr, Ala, Arg and Gly.
There should be a total of 16 (2×8) tRNAa.

4th One tRNA for each upper box of three.
So for Ile, there should be one tRNA.

5th One tRNA for each box of one (lower) So for Met and Trp there should be a total of two tRNAs.

In fact, we have seen that there must be two (and not one) tRNAs to recognise Met (one for the Mets incorporated in the process of the chain, the other for the initial Met which initiates the whole new peptide chain). These two tRNAs will have the same AUG anticodon but will otherwise differ.

A total of 32 different tRNAs is thus obtained.

It can be seen that in this wobble hypothesis the presence of A is excluded in the wobble position. In fact no upper box of one exists. Only the presence of an upper box of one (having a codon . . U) would require the existence of an anticodon having A in the wobble position.

The wobble hypothesis, according to which 61 tRNAs would not be needed to recognise 61 anticodons, but where, however, there would have to be a minimum of 32 tRNAs, has not been disproved by any trials. Nevertheless, work carried out in 1982 on yeasts seems to indicate that, in the well-defined case of the yeast *Saccharomyces cerevisiae,* I in the wobble position does not recognise A, and U in the wobble position does not recognise G, which would take to about 46 the number of tRNAs in this yeast. It remains no less true that a tRNA can recognise several codons. It can at present be concluded that the number of tRNAs needed to transport 20 amino acids must be between 32 and 61.

(b) In mitochondria

- Reminder of the genetic code in mitochondria

We recall that the proteins of mitochondria have a dual origin. The majority are coded by the DNA of the nucleus, synthesised in the cytoplasm and transported in the mitochondria, but some (proteins forming part of the internal membrane and playing an essential role in oxidative phosphorylation) are coded by the mitochondrial DNA (the mitochondria possessing its own machinery to synthesise proteins).

As we have seen, the genetic code of mitochondrial DNA is slightly different from the genetic code used in the cytoplasm (see Table 2):

— Ile has only one box of two (instead of one box of three): AUU and AUC (AUA here codes not for Ile but for Met!);
— Met has one box of two (instead of one box of one): AUA (which in the cytoplasmic system, codes for Ile) and AUG;

Table 2 — Genetic code (human mitochondria)

First nucleotide	Second nucleotide				Third nucleotide
	U	C	A	G	
	Phe	Ser	Tyr	Cys	U
	Phe	Ser	Tyr	Cys	C
U	Leu	Ser	Stop	~~Stop~~ Trp	A
	Leu	Ser	Stop	Trp	G
	Leu	Pro	His	Arg	U
	Leu	Pro	His	Arg	U
C	Leu	Pro	Gln	Arg	A
	Leu	Pro	Gln	Arg	G
	Ile	Thr	Asn	Ser	U
	Ile	Thr	Asn	Ser	C
A	~~Ile~~ Met	Thr	Lys	~~Arg~~ Stop	A
	Met	Thr	Lys	~~Arg~~ Stop	G
	Val	Ala	Asp	Gly	U
	Val	Ala	Asp	Gly	C
G	Val	Ala	Glu	Gly	A
	Val	Ala	Glu	Gly	G

— Trp has one box of two (instead of one box of one): UGA (which in the cytoplasmic system is a stop codon) and UGG;
— Arg has only four codons (CGU, CGC, CGA, CGG) instead of six (here, the two codons AGA and AGG are stop codons which do not code for any amino acids).

- U/N wobble

It has been shown that, in mitochondria, there are 22 tRNAs (considered so far to be the minimum required). This is explained by the 'U/N wobble' (which is only recognised in mitochondria) and which was described for the first time in 1980 by Sanger *et al.*

U in the wobble position on the anticodon can recognise N, i.e. any of the four bases.

- Thanks to the wobble, 22 tRNAs are adequate

In the case of protein syntheses occurring in mitochondria, all that is needed, therefore, is:

— a single tRNA to recognise the codons of a box of four (U/N wobble) for eight tRNAs (Leu, Val, Ser, Pro, Thr, Ala, Arg, Gly);

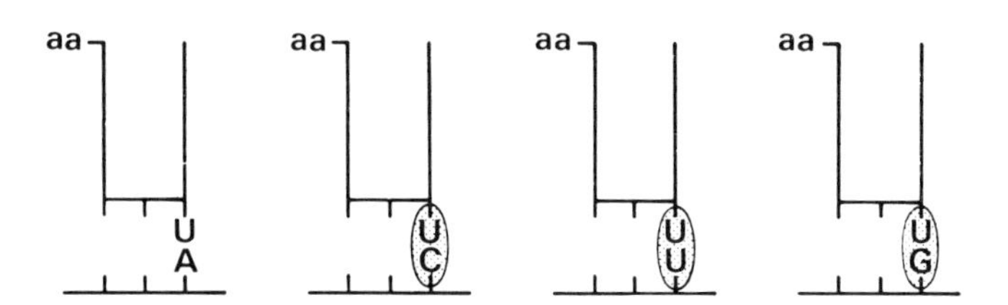

Fig. 73 — U/N wobble (human mitochondria).

— a single tRNA to recognise the codons of an upper box of two (G/U wobble) or eight tRNAs (Phe, Ile, Tyr, His, Asn, ASP, Cys, Ser);
— a single tRNA to recognise the codons of a lower box of two (U/G wobble) or six tRNAs (Leu, Met, Gln, Lys, Glu, Trp).

(As we have indicated, there are no longer any boxes of three (Ile) or of one (Met, Trp). These have become boxes of two.)

Contrary to cytoplasmic synthesis, a single tRNA is required for Met (whether it be the initial Met or the internal one).

Thus, the total number of tRNAs (all of which have now been identified) involved in protein synthesis in the mitochondria of mammals is 22, instead of a minimum 32 tRNAs needed for cytoplasmic protein synthesis in the same mammals.

3. Significance: what is the significance of these wobble linkages?

1st These wobble linkages allow the cell to make economies. In effect, thanks to the 'wobble' the cell does not need 61 different tRNAs in order to recognise 61 different codons, (or 60 tRNAs for 60 different codons in the case of mitochondria).

2nd Also, it is supposed that the base in the wobble position accelerates the speed of translation. In fact, the anticodon must not ally itself too firmly with the codon, otherwise the dissociation of the codon–anticodon complex would be too slow. The non-classic linkages allow a greater freedom of orientation. There is some 'play' or 'wobble' in the codon–anticodon linkage which is thus relatively slack.

H. Post-translational modifications

After translation, various types of modification may be produced.

1. Examples of permanent modification

Certain modifications are permanent, as for example:

- Shortening of the peptide chain: this is most frequently the way in which the initial methionine will be cleaved. In fact, none of the known natural proteins begin with methionine (or formylmethionine in the case of *E. coli*). After synthesis of the peptide chain, one or more amino acids are effectively cut off at the NH_2 end.
- Addition of sugars to the majority of the proteins secreted: the sugars hook most often onto the Asn by a bond called '*N*-osidic' with the NH_2 of the Asn residue

(and thus the bonds with the OH group of Ser or Thr are called '*O*-osidic'). This glycosylation takes place in the endoplasmic reticulum (ER) thanks to enzymes located on the lumen side of the ER membrane. In the Golgi apparatus, some sugars are removed. Other complex sugars can be added.

- Covalent addition of a coenzyme
- Hydroxylation of the Pro in the special case of collagen etc.

2. Examples of reversible modification

Other modifications are reversible, as for example:
- acetylation of the Lys, or
- phosphorylation of the Ser (this type of modification occurs in the histones).

I. Signal sequence

1. Which proteins have a signal sequence?

The proteins synthesised by a cell have several possible destinations. They can either:

— remain in the cytosol,
— be integrated into the plasma membrane, and also into the membranes of different intra-cellular organelles (in the case of eukaryotes), or
— be secreted outside the cell

- The proteins destined to remain in the cytosol will be synthesised by ribosomes said to be 'free'. 'Free' signifies that the ribosomes will not be linked to a cellular membrane (but that certainly does not prevent them from being linked to each other by one mRNA to form polysomes).
- The secreted proteins as well as the membrane proteins will be synthesised by ribosomes linked to the membrane of the ER. In the case of prokaryotes, the secreted proteins will be synthesised by ribosomes linked to the plasma membrane.

2. Synthesis of signal sequence proteins by linked ribosomes

A protein destined to be secreted enters into the aperture of the ER at the beginning of its synthesis. Here a problem arises. How can a protein in the course of synthesis cross the lipid membrane of the ER? Every protein destined not to remain in the cytosol possesses at its NH_2 terminal a short sequence of amino acids which are called 'leaders' or the 'signal sequence'. This signal sequence has the characteristic of being rich in hydrophobic amino acids. It is this which is responsible for attaching the ribosome onto the membrane of the ER, and for the penetration of the peptide chain into the vesicle of the ER.

3. The role of the SRP ('signal recognition particle')

More precisely in fact there exists in the cytosol a particle (comprising six peptide components +1 RNA) called the 'SRP' (signal recognition particle). It allies itself with the ribosomes soon after they have synthesised the signal sequence, and momentarily interrupts translation. This SRP particle also recognises a protein receptor situated on the membrane of the ER. This therefore facilitates the fixation of the ribosomes on the ER. The chain being produced (having the hydrophobic signal sequence at its NH_2 end) is then translocated across the double lipid layer while elongation restarts. This receptor protein is also humourously known as the 'docking protein': literally a protein which allows a ship (the signal sequence) to enter a dock (the vesicle of the ER).

To sum up

The whole of the onset of the synthesis of a protein destined to be exported from the cytoplasm takes place on a free ribosome. Subsequently, synthesis follows on the same ribosome which has become linked to the ER membrane of eukaryotes or to the plasma membrane in the case of prokaryotes.

The fact that a ribosome may be free or linked, therefore depends solely on the nature of the protein which it is about to synthesise. The signal sequence is a kind of 'postal code' which indicates that the protein is destined for export.

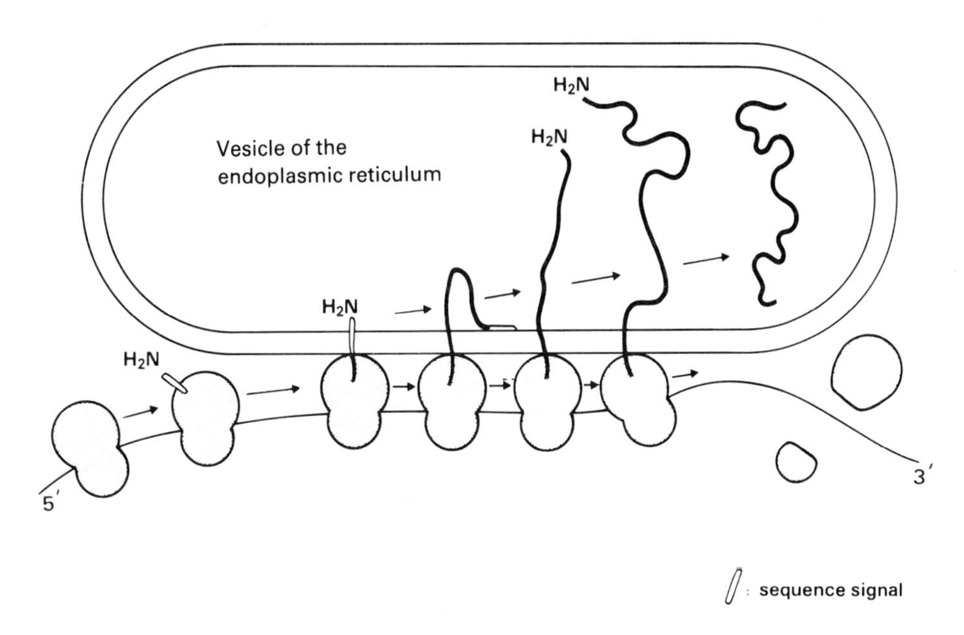

Fig. 74 — Synthesis of proteins destined not to remain in the cytosol.

4. *Cleavage of the signal sequence*

The signal sequence will not appear in the exported protein. In fact, it is cleaved as soon as the peptide chain has crossed the ER membrane (by an enzyme situated on the inner side of the membrane) even before synthesis of the peptide chain has ended.

5. *An experiment demonstrating the role of the signal sequence*

An interesting experiment was carried out in bacteria to demonstrate the role of this signal sequence. A coding gene was chosen for a protein which normally remains in the cytoplasm, for example: the gene for β-galactosidase. Another piece of coding gene is grafted on upstream of this gene, coding for an exported gene. Under these conditions the β-galactosidase will be synthesised as a result of the signal sequence and this β-galactosidase will then finally be found on the outside of the bacteria.

J. Some examples of biosynthesis

1. Insulin synthesis

Preproinsulin, proinsulin and insulin:

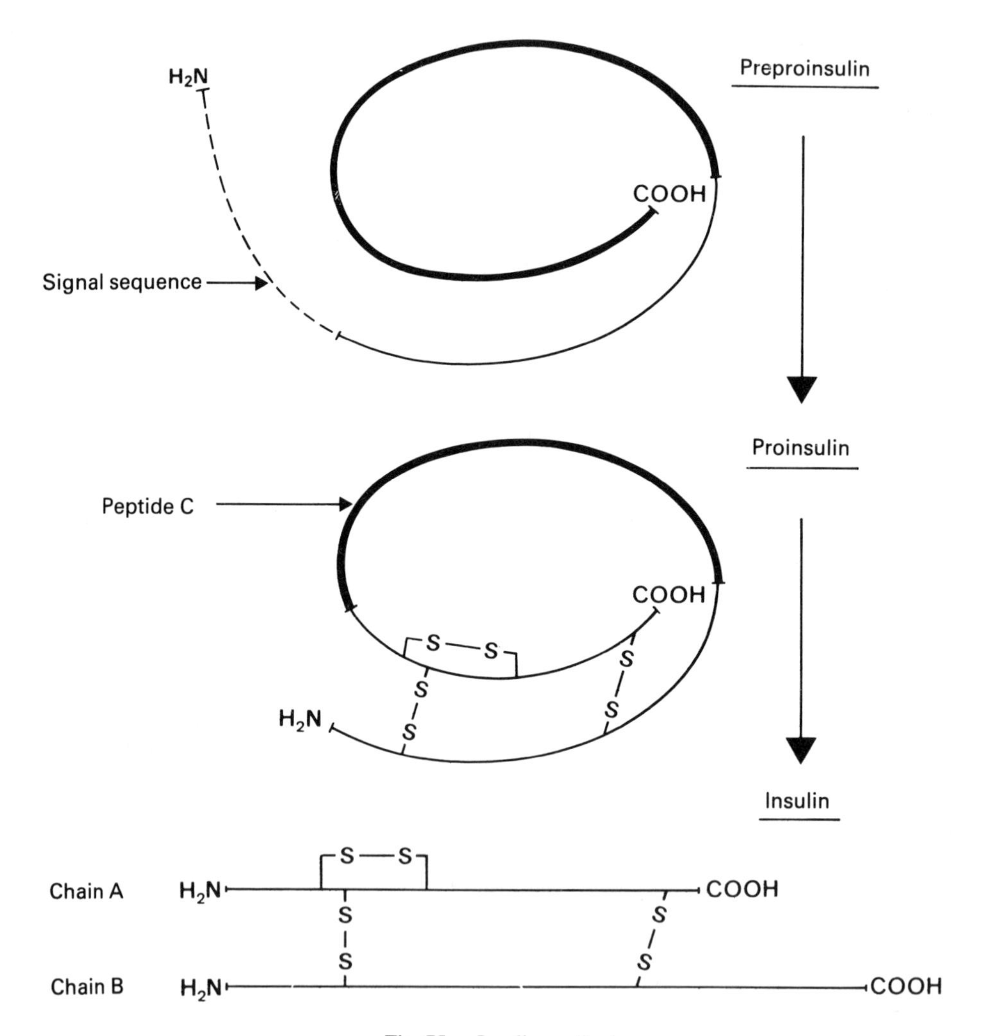

Fig. 75 — Insulin synthesis.

Insulin is a peptide hormone which plays an important role in the metabolism of sugars and which is missing in certain types of diabetes. It is synthesised by cells situated in the pancreas and is destined, as all hormones are, to be secreted by the cell which has synthesised it. In line with what we have seen above, insulin should then be synthesised with a single sequence. This is indeed the case, but in fact insulin is synthesised in an inactive precursor form and the signal sequence is found at the NH_2 extremity of this precursor. The peptide chain containing the signal sequence is called preproinsulin. This signal sequence will be cleaved in the ER to give proinsulin. An internal segment called peptide C (C for 'connection') will then be cleaved to finally give active insulin formed from two peptide chains linked by disulphide bridges. (Remember in this context that insulin was the first peptide whose sequence was determined, which incidentally won Sanger his first Nobel prize for chemistry in 1958).

2. *Immunoglobulin synthesis*

(a) *The four immunoglobulin chains*

Man and animals fight against foreign proteins (e.g. bacterial toxins) by manufacturing antibodies.

- The *antigen* is the foreign substance (generally a protein).
- the *antibody* is the protein substance synthesised by the organism to attack the antigen.

The antibodies are very numerous and specific for the different antigens encountered. There are millions of different antibodies and we only possess about 50 000 genes (to code for all the proteins of our organism. We are only just beginning to understand how our cells can produce millions of antibodies which are all different, with many fewer genes than the number theoretically necessary. It is an extremely complex phenomenon. Let us try to simplify it as much as possible.

The antibodies or 'immunoglobulins' (Ig) comprise essentially the IgMs, IgGs and IgAs (whose role is to neutralise the antigens), as well as IgDs (whose exact function is less well known), and IgEs (which play a role in allergic reactions).

Take the case of the IgGs. They are made up of four chains: two heavy specific chains, gamma (γ) chains (the heavy chains of the IgMs, IgDs, and IgEs are called respectively μ, α, δ, and ε) and two light chains (of lesser molecular weight), kappa (κ) or lambda (λ), which are also found in the other immunoglobulins. The part situated on the terminal NH_2 side (about 100 amino acids) is said to be 'variable' and the part situated on the terminal COOH side 'constant'. It is the part situated at the NH_2 end, the variable portion, which permits the specific association with the antigen.

(b) *Synthesis of a heavy mu chain (μ)*

In man (adult), the lymphocytes develop from stem cells contained in the bone marrow. These lymphocytes migrate towards the peripheral lymphoid tissues (spleen, lymphatic ganglia etc.). They become differentiated into B lymphocytes — immunocompetent cells capable of synthesising Igs. The first Ig to be synthesised by B lymphocytes is IgM. The basic unit of these IgMs is formed from two heavy mu chains and from two light kappa or lambda chains, thus: $\kappa_2\,\mu_2$ or $\lambda_2\,\mu_2$. (the IgMs will be the product of the association of five basic units). It is the heavy mu chain which will be synthesised first (before the light chains). Let us detail the mu chain synthesis a little more.

Recent discoveries using genetic engineering techniques have demonstrated that the coding sequences of DNA are not used in the same way in an embryonic cell and in an immunocompetent cell. We shall see that the structure of DNA is not 'fixed' as has been believed for years, but that nucleotide sequences initially distant from each other on this DNA can be brought together!

- An embryonic cell: sequence v′, D, J, C (distant)

On the DNA of an embryonic cell (or a cell other than an immunocompetent one) there are found:

— sequences which code for the variable portions of the heavy mu chain:
v′ sequences ('v'' for variable). Several of these exist, let us say 100 for example,
D sequences ('D' for diversity). Several of these exist, let us say 50 for example,
J sequences ('J' for junction). Several of these exist too, e.g. 6.

These figures should not be taken too literally. They are simply given as indicators to illustrate what is called a 'combinatory association' which will be described presently.

— a C sequence which codes for the constant portion of the mu chain. Sequences v′, D and J are far removed from sequence C.

- An immunocompetent cell: reshaping of DNA

In the cell, now capable of synthesising the Igs (B lymphocytes), a rearrangement of DNA sequences takes place:

one v′ sequence (and only one at random amongst the 100),
one D sequence (and only one at random amongst the 50), and
one J sequence (and only one at random amongst the 6)

are brought close to the C sequence. They are not in fact contiguous with sequence C, but separated by one intron.

In the course of the bringing together of v′, D, and J, the superfluous nucleotides will be eliminated. By contrast, the portion of DNA which separates the J sequence from the C sequence is not affected during the rearrangement. It will behave as an intron (and its transcription will be eliminated with the maturation of the mRNA).

Finally translation will give the heavy mu chain.

To simplify, we have not mentioned the signal sequence which precedes the v′ sequence (and which is separated from it by one intron), or the 'enhancer' sequences which are found in the vicinity of immunoglobulin genes (and which we discuss later).

In the example we have used, there are 100 v′ sequences, 50 D sequences and six J sequences, resulting in 100×50×6=30 000 possible combinations. This means that there are 30 000 different B lymphocytes, each capable of synthesising a different chain, giving a total of 30 000 heavy μ chains (differing in their variable portion).

It is now understood that to synthesise thes 30 000 different chains, 30 000 different genes are not necessary, but only 100×50×6+1=157 different sequences.

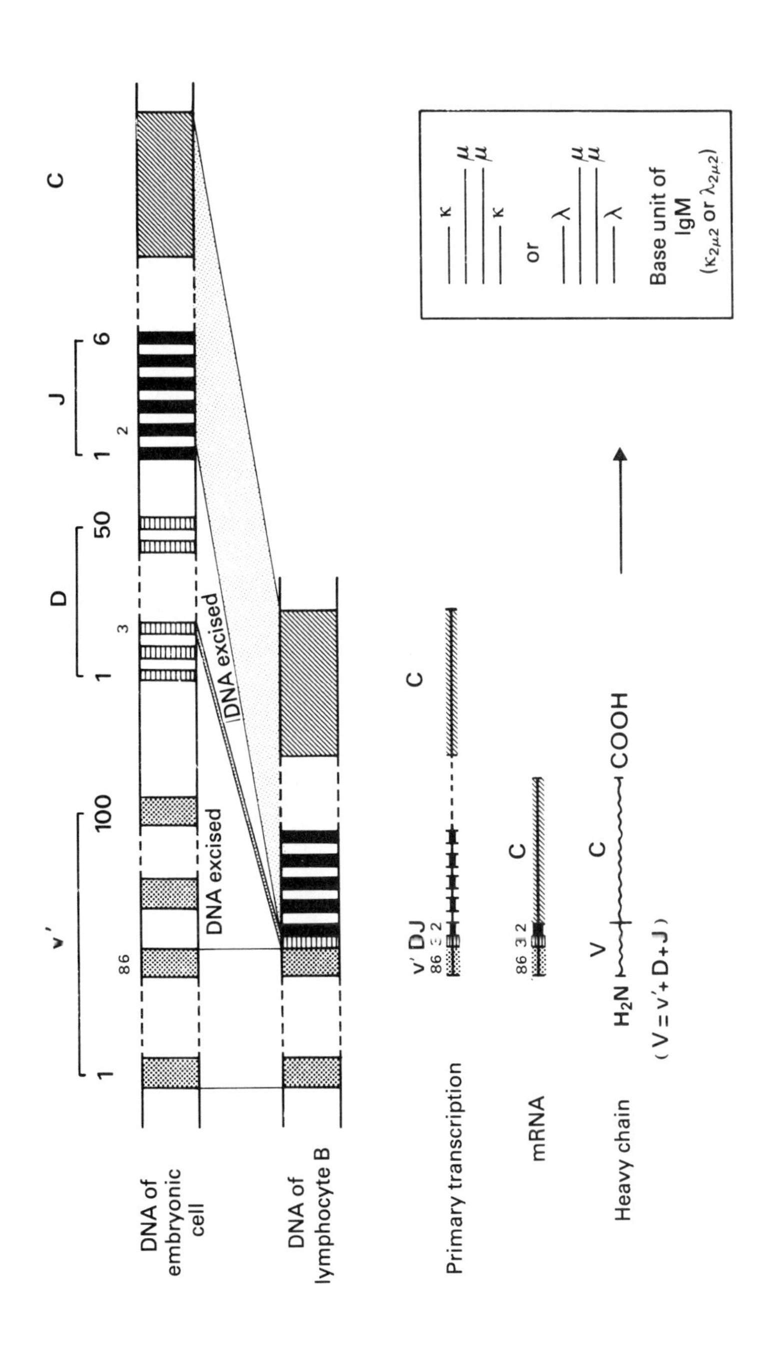

Fig. 76 — Synthesis of a heavy mu chain (example of the closing up of v′86, D3, J2.

The 'combinatory association', i.e. the random association of one of the 100 v' sequences with one of the 50 D sequences and one of the six J sequences has thus given 30 000 possibilities.

The number of possibilities is, in fact, even greater, because of the variation introduced by what is called the v'/D/J recombination flexibility. In effect, when a v' sequence associates with a D sequence, the splicing point (the point where the v' and D sequences will reunite) can vary. The same applies to the D and J sequences. Let us take the number of possibilities for heavy chains to about six times as many, say to 180 000. (We shall see next, with the light chains, an example of this recombination flexibility.)

Next it is the light chains which undergo a rearrangement of DNA in a comparable fashion.

(c) Synthesis of a light kappa chain (κ)

The principle of rearrangement of DNA sequences for light chains is comparable to

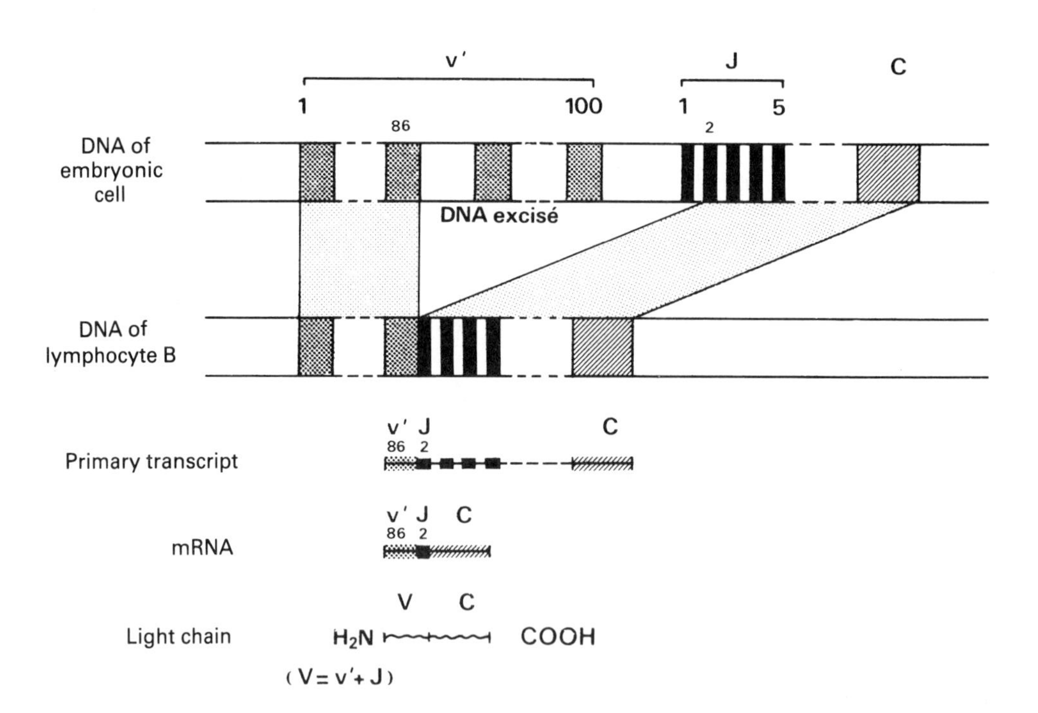

Fig. 77 — Synthesis of a light kappa chain (example of the closing up of v'86, J2.

that described for the heavy chains, but, in the case of light chains, there are only two types of sequence coding for the variable portion. These are the v' sequences (say about 100) and the J sequences (say about five).

Thus, 500 different combinations are obtained. There too, because of recombination flexibility (v′/J) the possibilities are again increased.

Here now, with regard to light kappa chains is the example promised to illustrate combination flexibility. To simplify, a single strand of DNA is represented, the strand of DNA complementary to the transcribed strand of DNA. (Remember that the strand not transcribed is identical in polarity and in bases to the mRNA — with the exception of Ts which will be replaced by Us in the mRNA).

TCTCCTCC	GTGGACG
1234	1234
Side 3′ of sequence v′	Side 5′ of sequence J

Splicing can take place (indicated by /):

at 1–1	TCT	CC/G	TGG	ACG
	Ser	Pro	Trp	Thr
or at 2–2	TCT	CCT	TGG	ACG
	Ser	Pro	Trp	Thr
or at 3–3	TCT	CCT	C/GG	ACG
	Ser	Pro	Arg	Thr
or at 4–4	TCT	CCT	CC/G	ACG
	Ser	Pro	Pro	Thr

In this example therefore a variation factor of three has been introduced (the amino acid can be Trp, Arg or Pro according to splicing), which will thus increase the number of combinations, which now rises to 1500.

In order to simplify we shall not discuss the rearrangements of λ chains. The principle is comparable to that of the κ chains.

(d) Association of heavy chains and light chains

Because of what is called 'isotopic exclusion' a cell will only synthesise (in addition to heavy chains) kappa chains or lambda chains. The heavy chain–light chain assemblage will thus end in the formation of immunoglobulins having two heavy chains and two light chains, with either two kappa or two lambda chains — not one kappa and one lambda chain.

The association of two heavy mu chains (identical) two light chains (identical), kappa for example, takes place in the cytoplasm of the cell.

This association further increases the number of possibilities. Thus, continuing the examples previously chosen, one obtains:

$$(100\times50\times6\times6)\times(100\times5\times3)=180\,000\times1500,$$

say about 270 million different IgMs

Other sources of variation exist, such as the point mutations which can occur in the variable zone and will further increase the total number of different possibilities.

(e) Localisation of the synthesis of heavy and light chains on the chromosomes (14,2,22)

It must be pointed out that the syntheses of light and heavy chains are effected in a cell by different chromosomes. Thus in man:

- chromosome 14 synthesises heavy chains ($\mu,\delta,\gamma,\varepsilon,\alpha$)
- chromosome 2 synthesises light kappa chains
- chromosome 22 synthesises light lambda chains

In an immunocompetent cell the 'v′DJ recombination' for the heavy chains or v′J combination for the light chains only occurs on one of the two homologous chromosomes. Only the gene coming from the conveniently rearranged DNA will be expressed (on only one of the two parent chromosomes). This is called 'allelic' exclusion. It is an exceptional case! For the other proteins coded by autosomic genes, it appears that the maternal and paternal genes are expressed in roughly equal quantities.

(f) Membranous IgM — secreted IgM commutation

According to the theory of selectional cloning presently accepted, differentiation has yielded millions of B pre-lymphocytes, differing in the membranous IgMs (and IgDs) carried on their surface. When a specific antigen is recognised by a surface IgM, the carrier B pre-lymphocyte of this IgM is activated. It then multiplies and forms a clone of B lymphocytes capable of synthesising great quantities of antibodies which have the same site for linking with the antigen (and so recognising the same antigen). When this activation or maturation occurs, the IgMs (and IgDs) disappear from the surface of the lymphocytes and the latter now start to synthesise the IgMs (as well as the IgGs, IgEs and IgAs) destined to be secreted. The two types of IgM differ at the COOH extremity of the mu chain, which is hydrophobic in the case of membranous IgMs (so it can establish itself in the lipophile membrane), and hydrophilic in the case of secreted IgMs. This difference between the two forms of IgM is created at the time of transcription. The same segment of DNA containing nucleotides coding for the hydrophilic sequence, intron and hydrophobic sequence, will be transcribed in two different ways.

- Formation of a long transcript (from 0 to 'a')

The primary transcript is then 'treated' (or 'processed'): the transcript of the intron (which comprises the hydrophilic sequence) is excised.

Translation finally gives the IgM with the hydrophobic sequence at the COOH end.

- Formation of a short transcription (from 0 to 'b')

This transcription contains the start of the transcription of the intron. As it does not contain the signal 'end of intron transcription', there will be no excision of this half of the intron transcription (rich in hydrophilic amino acids). The translation will finally give an IgM identical to the preceding one, except that the sequence at the COOH extremity is this time hydrophilic, and will allow secretion of this IgM.

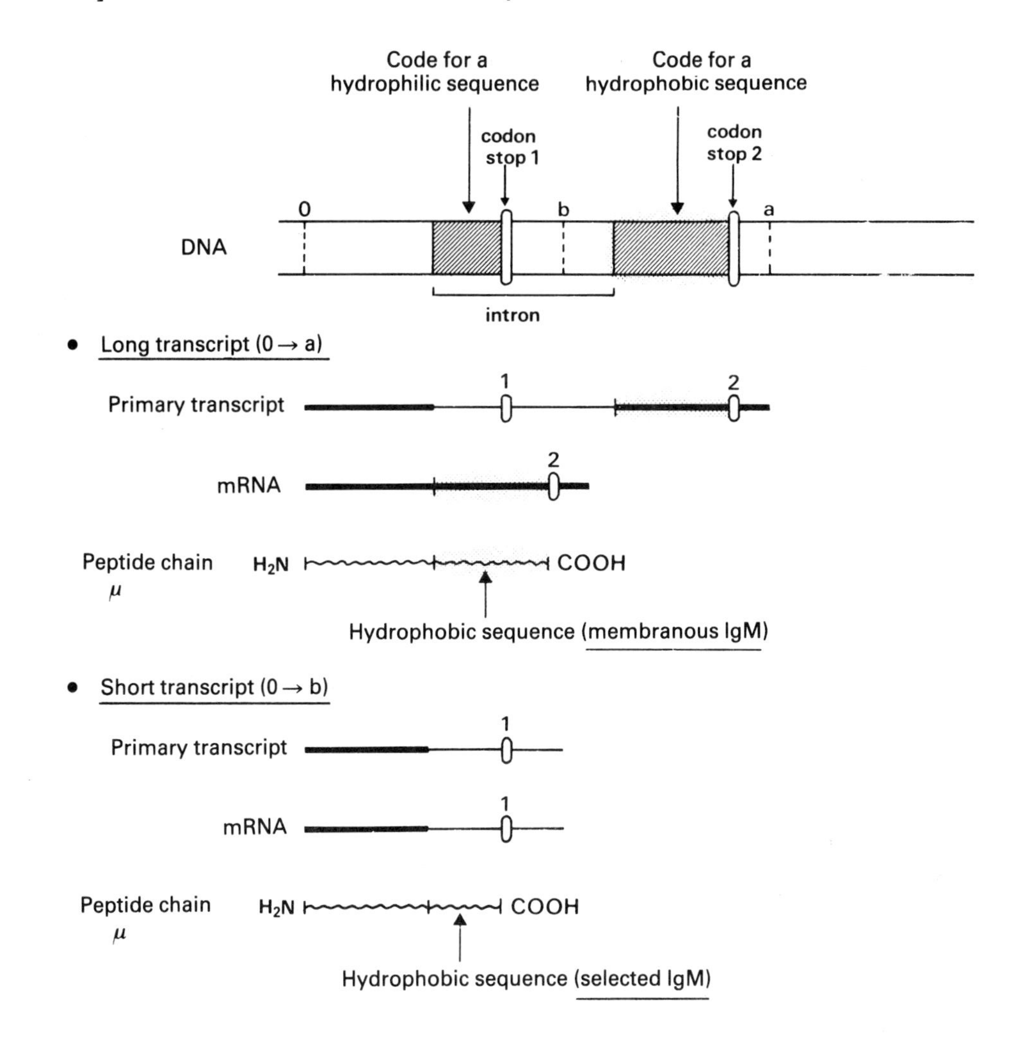

Fig. 78 — Membranous IgM — secreted IgM commutation.

(g) Commutation of heavy ('switch') chains

The IgM commutation to the other types of Ig (IgG, IgE, IgA) must now be explained. In fact, the B lymphocyte which began by synthesising the IgMs will then synthesise new types of antibodies having the same light chains as well as the same variable portion of the heavy chains. They will therefore have the same bonding site for the antigen. What will be different in each class of Ig (and so be characteristic of the Ig class) is the constant part of these heavy chains. The lymphocyte will also synthesise, for example, gamma chains (IGC) or alpha chains (IgA), instead of by mu chains (the IgMs).

This commutation is explained by the fact that the DNA sequence coding for the constant part of a heavy chain in fact contains several constant sequences arranged in this order:

mu, delta, gamma 3, gamma 1, gamma 2b, gamma 2a, epsilon, alpha

The gene, having undergone v′DJ rearrangement, is represented in Fig. 79(a) before change of class. Fig. 79(b) shows an example of IgM commutation IgG3 (the segment coming between J and gamma 3 is excised).

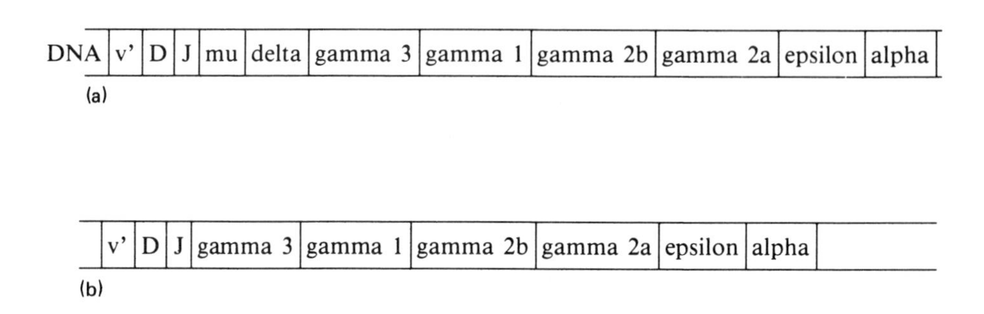

Fig. 79 — Commutation of heavy immunoglobulin chains.

When a B lymphocyte secretes IgGs and then IgEs or IgAs, it will use the same variable sequences rearranged, and it will associate them with other constant characteristic segments of the antibody class. The constant sequences situated upstream of the constant sequence expressed will be eliminated. Thus, in a B lymphocyte producing gamma chains, a deletion of mu and delta sequences will be observed.

(h) The signals implicated in rearrangements of DNA and commutations

The question to be asked now is obvious: How are the DNA sequences brought together at the time of v′/D/J or v′/J rearrangements, or of commutation of heavy chains?

As far as the v′/J rearrangement is concerned, recombination signals are identifiable below the v′ (at 3′) sequence and above the J (at 5′) sequence. Enzymes (still unknown) would cut at this point, eliminating the recombination sequences and splicing the 3′ extremity of a v′ sequence with the 5′ extremity of a J sequence. Signals of the same type are equally identifiable above and below D sequences, thus explaining the v′DJ rearrangements of heavy chains.

The signals for IgM commutations to the other Igs, are less well known.

(i) The rearrangement at mu precedes that at kappa which itself precedes that at lambda

The study of different types of leukaemias where the B lymphocytes are ‘frozen’ at different stages of development, has prompted the suggestion of a hierarchical order in the rearrangement of DNA sequences: rearrangement at mu would precede that of light chains, and rearrangement at kappa would precede that at lambda.

When a B lymphocyte secretes kappa light chain Igs, the DNA sequences corresponding to lambda light chains are never rearranged and remain in the embryonic state. Conversely, when a lymphocyte synthesises lambda light chains, the sequences corresponding to kappa chains are deleted or rearranged in an aberrant way.

Several teams of researchers have recently envisaged the possibility of a cell somehow possessing four chances of making a correct B lymphocyte. They suggest that the cell always begins with a kappa rearrangement. If this is prevented a second attempt is made with the second kappa allele. If this is also prevented an attempt is made to use a lambda allele. If this is once again blocked, and if a rearrangement with the second lambda allele also leads to a block, the cell has wasted its four chances. It will be immunologically incompetent and will probably not survive.

In order to create a light chain the cell would also make use of 'three spare wheels'.

3

Regulation of protein synthesis

The mechanisms which control the quantity of proteins synthesised are actually much better known in prokaryotes.

I. REGULATION OF PROTEIN SYNTHESIS IN PROKARYOTES

For reasons of cellular economy it is desirable, as can well be appreciated, that mechanisms exist which, as far as the needs of the cell are concerned, are capable of regulating the synthesis of proteins:

— if there is need, synthesis must be effected,
— conversely, if the protein is present in excess, it must be possible to arrest synthesis.

A. Regulation during transcription

We shall see successively:

- induction, or how synthesis of a protein is initiated, and
- repression, or how synthesis of a protein is blocked.

The examples chosen will concern enzymatic proteins.

1. Induction — e.g. the lactose operon

(a) Experiments illustrating the functioning of lactose operon

If colibacilli are cultured on a medium containing glucose, the colibacilli will multiply without difficulty. If glucose is now suppressed and the bacteria are put on a lactose-containing medium, they no longer grow. It would seem just as though they were unable to utilise lactose. However, lactose is a saccharide which contains glucose (β-galactosidoglucose).

After a certain time, however, growth resumes. The colibacilli can now utilise lactose (to liberate the glucose) thanks to a β-galactosidase which they synthesise for the purpose. In fact, the utilisation of lactose requires three enzymes:

— a permease which facilitates the passage of lactose across the membrane of the bacterium,
— an acetylase whose physiological role is not well known,
— a β-galactosidase which permits, as we shall see, hydrolysis of lactose into galactose and glucose.

It is important to understand that these three enzymes are not always present in the colibacilli (except in very small amounts). They are only synthesised if needed to liberate glucose from the lactose. These three enzymes are said to have been 'induced' by the lactose and that the lactose is an 'inducter'.

(b) Description of the lactose operon
Jacob and Monod (Nobel prize for physiology and medicine 1965) provided the description of the lactose operon (see Fig. 80).

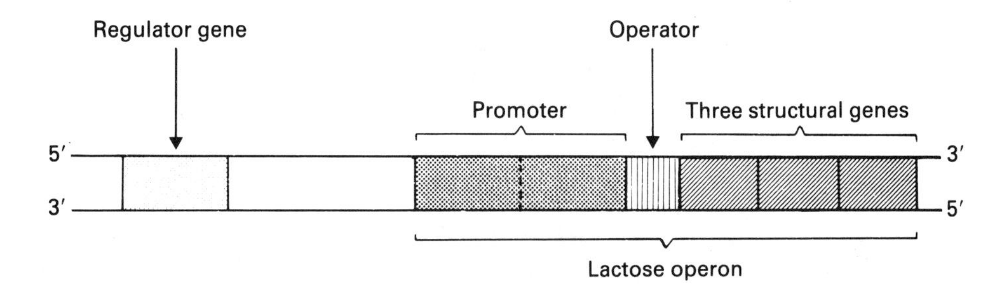

Fig. 80 — Description of lactose operon.

The lactose operon is the segment of DNA which contains the three genes coding for the three enzymes needed for the utilisation of lactose by the colibacillus. In addition to these three genes, called 'structural genes', the lactose operon contains what are termed:

— the promoter and
— the operator.

Upstream from the lactose operon is

— the regulator gene.

(c) Functioning of the lactose operon
Let us return to the previous experiments to explain them on the molecular scale:

- In the presence of glucose (see Fig. 81).

The three genes coding for the three enzymes destined to use lactose will not be expressed (which is entirely logical, since glucose is present in the medium, and especially as this medium is deprived of lactose, the bacterium has no need to manufacture enzymes which utilise lactose). It is said that there is 'repression' (or that the 'genes are repressed', or that 'lactose operon is repressed', or that 'synthesis is repressed').

How does this repression come about? The regulator gene codes for a protein which is called a 'repressor'. The repressor can recognise and block the operator. If a repressor is fixed on the operator, it sterically impedes the attachment of the RNA polymerase to the promoter. There is therefore no transcription (no mRNA, and no synthesis of lactose-utilising enzymes).

To sum up: the regulator gene has synthesised a represser which blocks the operator. No transcription is possible.

- In the presence of lactose (see Fig. 82).

In this case the medium is deprived of glucose but contains lactose. If it wishes to survive the bacterium will have to utilise the lactose. For the bacterium there is therefore a vital need to synthesise enzymes capable of hydrolysing the lactose into glucose and galactose.

In the presence of lactose (i.e. the inducter) repression is lifted, and the genes can then be expressed: there is said to be a derepression.

The repressor synthesised by the regulator gene is recognised by the inductor (i.e. by the lactose) and associates with it. The repressor then becomes incapable of recognising the operator and of becoming attached to it. If the repressor is already fixed on the operator it will be unhooked by the inducter.

The RNA polymerase can now become fixed on the promoter. There is no longer steric impedance and transcription begins. An mRNA is formed and the three enzymes needed for utilisation of the lactose will be able to be synthesised. (In this regard it must be noted that in prokaryotes, as previously mentioned, an mRNA strand can be polycistronic — it can contain the information necessary to form several different proteins — which is not the case in eukaryotes where one mRNA strand only codes for one kind of protein. In suppressing the block caused by the repressor, the lactose has allowed the synthesis of the three enzymes needed for its utilisation. It is now known that lactose itself is not the true inducer: it first has to be transformed into a related compound, allolactose.

Summary

Less than ten molecules of β-galactosidase per colibacillus are synthesised in the presence of glucose.

Several thousand molecules of β-galactosidase are, however, synthesised in the presence of lactose (and in the absence of glucose). This can be summed up in three words; there has been: '*induction by derepression*'.

In fact we are dealing with a negative regulation:

— the repressor prevents formation of mRNA

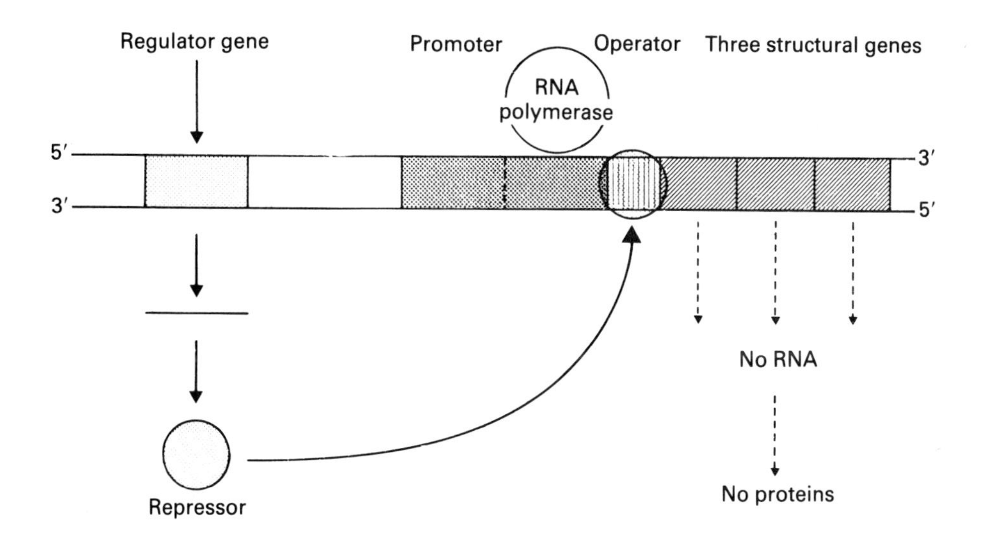

Fig. 81 — Functioning of the lactose operon in the presence of glucose.

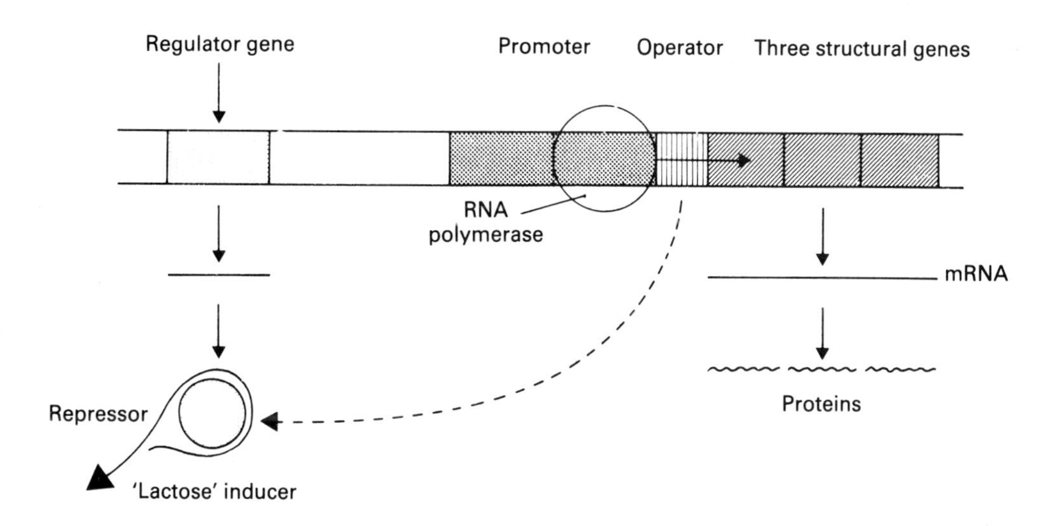

Fig. 82 — Functioning of the lactose operon in the presence of lactose.

— the inducter prevents the preventer from preventing.

It is also known that there is another kind of regulation, positive this time. Let us return to our experiment of culturing colibacilli, but now in the presence of glucose + lactose.

- In the presence of a mixture of glucose and lactose: regulation is positive (cAMP, and the CAP protein are involved) (see Fig. 83).

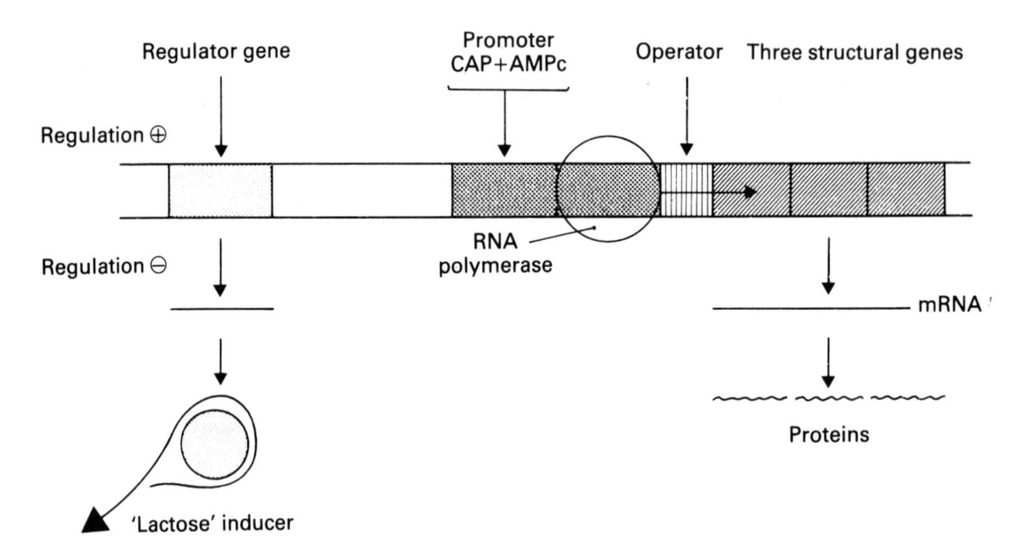

Fig. 83 — Functioning of lactose operon in presence of a glucose + lactose mixture.

In the presence of a glucose + lactose mixture, the bacterium firstly chooses to metabolise the glucose and when the glucose is exhausted, it will utilise the lactose.

In this type of positive regulation, one substance plays a very important role, namely cyclic AMP (cAMP).

It was a great surprise when cAMP was found in a colibacillus. This substance was known of in mammals where it plays the role of 'second messenger' to permit the action of certain hormones.

In a colibacillus, when the quantity of metabolisable glucose diminishes (for example because the glucose of the culture medium utilised by this colibacillus has not been renewed and so is exhausted), cAMP increases. This is a kind of signal signifying that the bacterium 'is hungry'. This cAMP thus has a positive effect on the lactose operon: it accelerates the synthesis of enzymes permitting lactose to be used.

First of all cAMP fixes itself onto a protein which is called CAP 'catabolism gene activator protein', a protein which activates a catabolic gene, (not to be confused with the 'cap' of the mRNA of eukaryotes). The cAMP–CAP complex then attaches to the promoter in a neighbouring site where the RNA polymerase is fixed. Once fixed, this complex increases the affinity of the RNA polymerase for the promoter. The transcription can thus be multiplied by a factor of 50.

2. Repression – e.g. the tryptophan operon

(a) *Bacterial synthesis of Trp*

Trp is an indispensable amino acid for man and must be supplied to us through nutrition. It is not the same for bacteria: the colibacillus is in fact perfectly capable of synthesising this amino acid.

To synthesise Trp a whole series of reactions has to be produced: an enzyme intervenes at each stage.

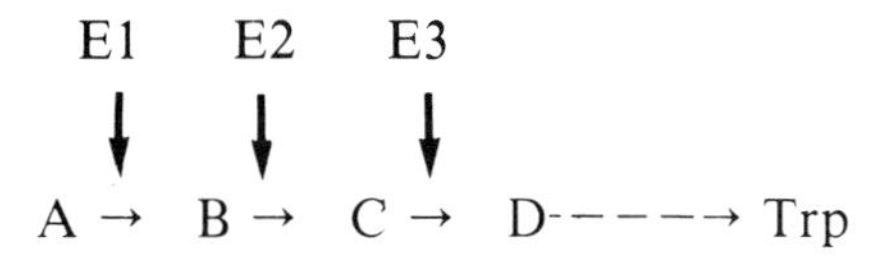

If these enzymes are themselves synthesised by the bacteria, the series of reactions culminating in Trp proceeds without problems, and Trp is finally obtained. However, if these enzymes are no longer synthesised, the production of Trp is obviously arrested.

We shall see that synthesis of the enzymes will be active when the cell lacks Trp. However, enzyme synthesis will be repressed when the cell has sufficient Trp.

(b) Functioning of the Trp operon

Let us create a diagram of the Trp operon (see Fig. 84), the collection of genes which code for the enzymes needed for the synthesis of Trp.

- In the absence of tryptophan

The regulator gene synthesises a repressor, said to be 'inactive', because it does not attach itself onto the operator. This is called an 'apo-repressor'. RNA polymerase can then attach itself on the promoter and initiate the transcription. The structure genes are transcribed, mRNA is formed, and the enzymes are synthesised. Trp can then be produced.

- In the presence of tryptophan

The addition of Trp to culture medium causes an abrupt arrest of Trp synthesis. (This is logical since there is Trp in the medium and the bacterium has no further need to synthesis any.) Synthesis is said to be repressed.

How does excess of Trp block its own synthesis? Trp, which in this case is called the 'co-repressor', combines with the inactive repressor, which then becomes capable of attaching to the operator, and blocks it. The RNA polymerase cannot initiate transcription and so transcription is prevented. It is said that there has been

1. In the absence of tryptophan:

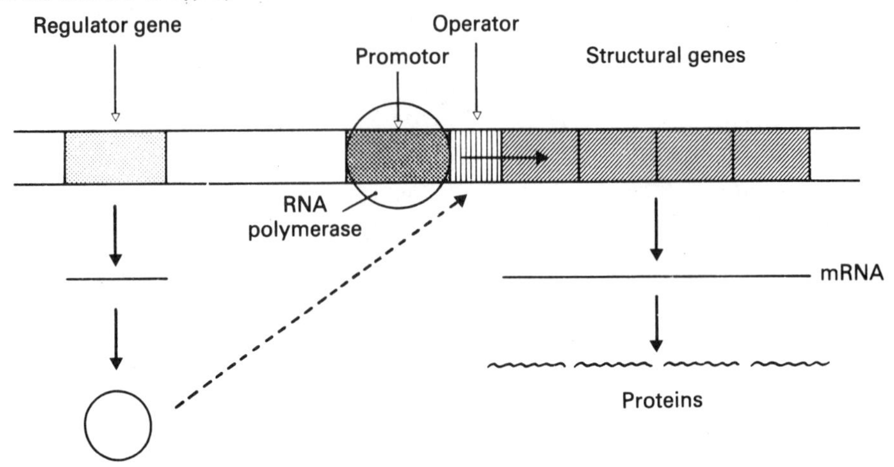

Inactive repressor = apo-repressor

2. In the presence of tryptophan:

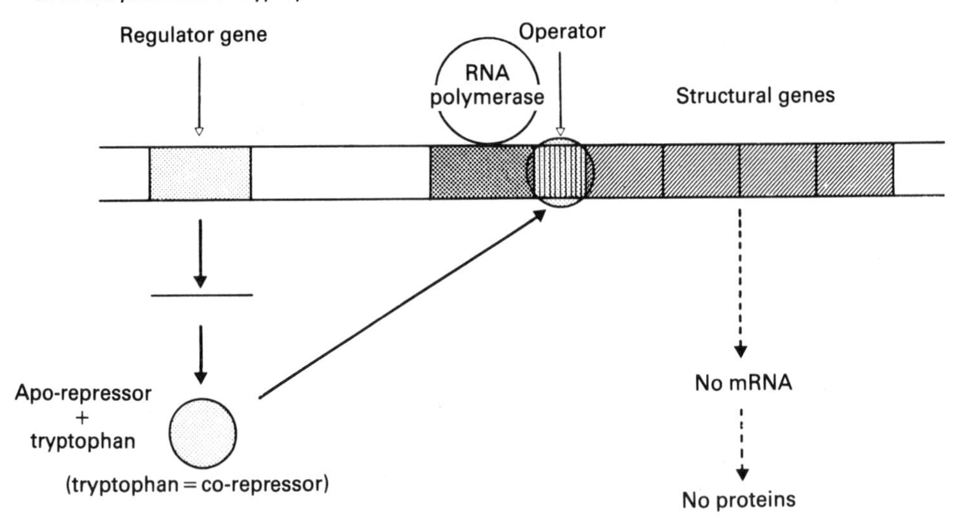

Fig. 84 — The tryptophan operon.

repression by the terminal metabolite (in this case the Trp) which has thus inhibited its own synthesis. This synthesis will be blocked until all the Trp has been utilised by the bacterium. When there is no more Trp in the medium synthesis restarts. To sum up:

— in the absence of Trp, transcription takes place,
— in the presence of Trp transcription is blocked.

The functioning of the Trp operon is in fact more complicated. Another site also intervenes in addition to the operator site, namely the attenuator site. However, in the interests of simplification we shall not mention it here.

3. Comparison of induction and repression

We have seen, in induction and repression, two different mechanisms by which the bacterial cell can regulate the synthesis of its proteins and so meet problems of economy. We have described:

(a) A permanent repression

Repression is permanent but may be lifted through the presence of an inducter. This is derepression by induction. (This sort of negative command is encountered in every-day life: 'taps' of water, gas, and electricity are usually closed and are opened only in case of need.)

This type of regulation is generally observed for enzymes which degrade (e.g. β-galactosidase).

(b) A permanent synthesis

Synthesis is permanent but may be interrupted by command of the terminal metabolite itself (this situation may be compared to that where water runs permanently into a basin and is stopped when the water reaches a certain level).

This type of regulation is usually observed for those enzymes which synthesise (e.g. the enzymes of Trp synthesis).

For clarity, in the previous diagrams the promoter was shown beside the operator. In fact, there is overlapping of the promoter and the operator.

B. Regulation during translation

E.g. Synthesis of r-proteins.

1. Significance

We now know of another type of regulation of protein synthesis relating to the process of translation. One of the best studied examples is that of the synthesis of ribosomal proteins (r-proteins) in *E.coli.*

We recall that ribosome of *E.coli* contains 52 different r-proteins coded for by 52 different genes, as well as three different rRNAs coded for by several series of three different genes. (These three different genes of rRNA will, for each series, be transcribed as a single rRNA precursor, then cut to form rRNA 16 S, rRNA 23 S and small rRNA 5 S.)

The making of a ribosome consumes a great deal of energy. In order to avoid any waste, ribosome production must be adjusted to cellular needs. For ribosomes assembly, regulation may be exercised during either transcription or translation. Let us consider only the second type of regulation, which will lead to the discovery of the 'translation repressors'.

2. r-Proteins in excess block their own mRNA

As long as rRNAs are available in the cell, synthesised r-proteins are fixed onto these rRNAs, ribosomes assembly follows its course and r-protein synthesis continues.

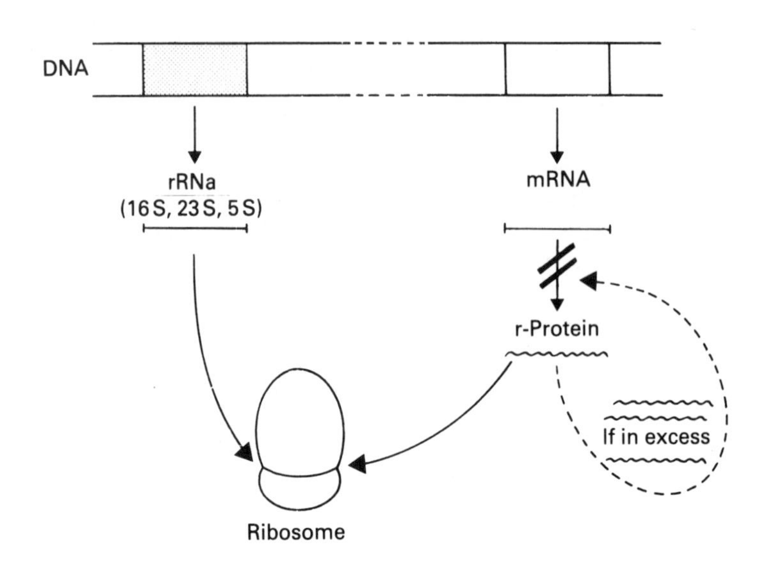

Fig. 85 — Regulation during translation: the example of the synthesis of r-proteins. For simplicity, only the blockage by an r-protein of its own synthesis has been represented here. (However, in certain cases, the r-protein can equally block the synthesis of other r-proteins produced by the same operon.)

When r-proteins are present in excess they will block their own mRNA. Thus, they act like translation repressors.

More precisely they block not only their own synthesis but also that of other r-proteins situated in the same operon. (In fact it is very complicated because not all the r-proteins of the operon are subjected to repression, only certain ones.)

It is interesting to note that an r-protein can thus attach itself either to an rRNA (to participate in the constitution of the ribosome) or to an mRNA (to block the translation of this mRNA).

3. *How is the affinity of r-proteins for both rRNAs and mRNAs at once explained on the molecular scale?*

It is now known that there is a structural resemblance between the rRNA site where the r-protein will be fixed, and the rRNA site where this r-protein can also be attached. These resemblances relate to the nucleotide sequence (which is, therefore, very similar for the two types of RNA at the site of fixation), and to the secondary structure (for at these sites the RNAs are folded to form a bicatenary area having a comparable shape). Thus, there is a site common to rRNAs and mRNAs for receiving the r-protein.

When rRNA and mRNA are both available, the r-protein fixes itself more easily to the rRNA than to the mRNA.

Since it is more rapid, translational control is more effective than transcriptional control, which can only take effect after degradation of pre-existing mRNAs.

C. How does protein–nucleic acid recognition occur?

1. Examples of protein–nucleic acid recognition

Recognition between a protein and a nucleic acid is a very remarkable phenomenon. In fact, these two types of molecule are very different. Whereas protein is composed of amino acids, nucleic acid consists of nucleotides (themselves formed by the base combination + sugar + phosphoric acid).

In the study of protein synthesis this type of recognition between a protein (usually an enzyme) and a nucleic acid is seen at every stage. For example:

- Aminoacyl–tRNA synthetase–tRNA.
- RNA polymerase–promoter.
- RNA polymerase–signals to end transcription on DNA.
- Repressor–operator.

2. Possible recognition mechanisms

Let us cite two mechanisms capable of playing a role in protein–nucleic acid recognition.

(a) Harmony between the symmetry of the protein and that of the nucleic acid

Take the case of repressor–operator recognition. The region of DNA where the operator is to be found is composed of nucleotides arranged with a certain amount of symmetry (an imperfect palindrome) which could be represented as a mesh with several 'holes'. The repressor, which is a protein possessing a certain symmetry, will recognise this specific sequence, as a setting with several pegs which are spatially complementary to it (i.e. into which it will fit) (see Fig. 86).

5' — TGTGTG G AATTGT G A G C G G A T A ACAATT T CACACA — 3'
3' — ACACAC C TTAACA C T C G C C T A T TGTTAA A GTGTGT — 5'

Fig. 86 — The operator (of lactose operon) and the 'gaps': region of imperfect symmetry. The symmetry of the repressor (a protein) must harmonise with that of the operator (DNA).

(b) Hydrogen bonds

Hydrogen bonds can intervene between an amino acid of the protein and a nucleic acid base. For example, hydrogen bonds can be found between an Asn (of a protein) and an adenine (of a nucleic acid).

What is particularly remarkable is that a protein should be capable of recognising a sequence of only a few nucleotides on a nucleic acid capable of being formed, in the case of DNA for example, from several million nucleotides.

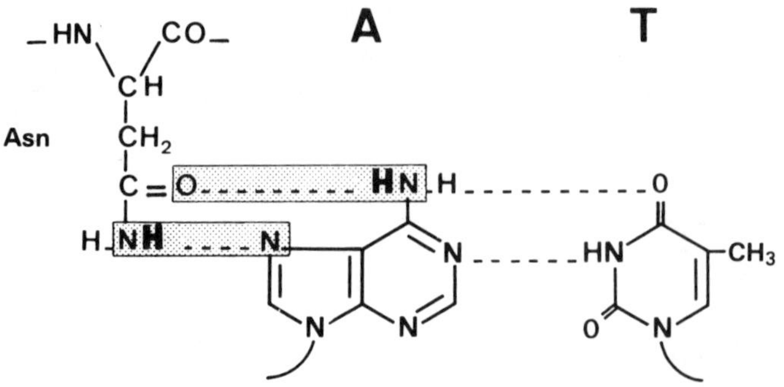

Fig. 87 — Example of hydrogen linkages between Asn (of a protein) and adenine (of a nucleic acid).

A final, philosophical remark. Proteins (in particular the enzymes mentioned above) are necessary for the synthesis of new molecules of DNA. It is the same for synthesis of other types of nucleic acid (such as rRNAs and tRNAs) where numerous enzymes (and thus proteins) intervene every time. Conversely throughout protein synthesis, nucleic acids (DNA, mRNA, rRNA, tRNA, snRNA) have been necessary to bring this synthesis to a successful conclusion. One can only restate this astonishing mystery: to synthesise proteins there must be nucleic acids, and to make nucleic acids there must be proteins. So, at the beginning of time, which were the first molecules: the DNA or the proteins? This recalls the well-known dilemma: 'which came first, the chicken or the egg?'

II. REGULATION OF PROTEIN SYNTHESIS IN EUKARYOTES

The regulation of protein synthesis in eukaryotes, and more particularly in man, is much more complex than in prokaryotes. This is not surprising when one considers that DNA is found here in a nucleus, that this DNA contains about one thousand times as many base-pairs and that the cells which contain this DNA are specialised.

A. Resumé of the structure of the eukaryote chromosome — nucleosomes, histone and 'non-histone' proteins

The chemical substance which constitutes the chromosomes is called chromatin. It is a mixture of DNA and proteins. Chromatin is composed of a collection of particles called 'nucleosomes' piled one on top of each other in a pattern known as a 'pearl necklace' (each pearl being a nucleosome).

A nucleosome may be represented as a cylinder made up of proteins. Around this 'bobbin' of proteins, the DNA entwines itself forming $1\frac{3}{4}$ of a left-hand super helix.

In this model the DNA filament which is entwined around a nucleosome also binds the other nucleosomes.

The proteins detected in the chromatin comprise notably:

- 'Histone' proteins

These constitute the heart of the nucleosome. These are basic proteins rich in amino acids such as arginine and lysine. They are thus rich in positive charges which will

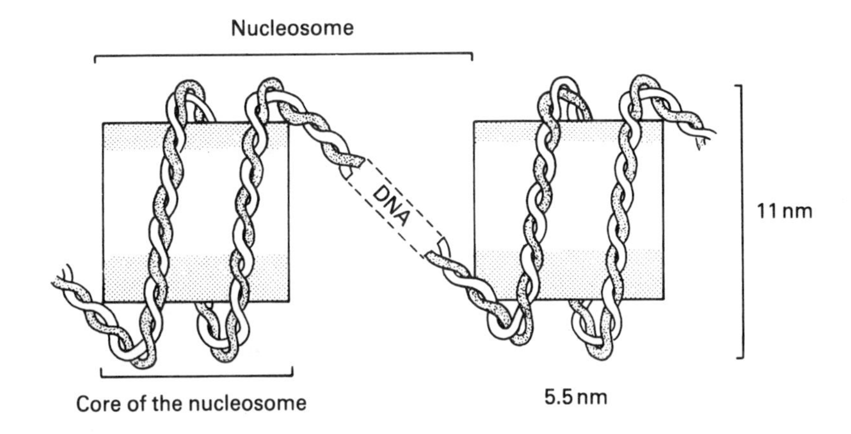

Fig. 88 — A nucleosome. (Bradbary, E. M. 'Chromatin'. *La Recherche*, July–August 1978, p. 650, Fig. 6B.

enable them to form ionic bonds with DNA (which carries negative charges due to the free acid functions of the phosphoric acid group).

These histones can undergo reversible covalent modifications, provoking changes of charge. Thus lysine can be acetylated, which suppresses positive charges. Conversely, serine residues can be phosphorylated, which this time elicits negative charges. This possibility for rapid and reversible change in the nature of chromatin charges certainly plays an important physiological role.

- Non-histone proteins

The majority of these proteins are only present in very minor amounts and are still not well understood.

The GRP group (gene regulatory proteins) is found here. These gene regulatory proteins are essentially activators in eukaryotes. They combine with DNA on a site adjacent to that of the RNA polymerase. They thus facilitate the binding of RNA polymerase and activate transcription. It is therefore a case of positive regulation (comparable to that seen in prokaryotes).

B. Gene expression

In the eukaryotes not all the structural genes of a cell are transcribed, i.e. not all the structural genes will give rise to the formation of mRNA and proteins. This is due to the fact that, in the higher eukaryotes, the cells are specialised. Thus nerve cells do not synthesise the same proteins as the muscle cells or liver cells. However, in theory they could do so, since all cells contain the same chromosomes and therefore the same DNA, and the same genes. It is said that not all the genes in a eukaryote will be expressed.

1. *Experiments involving transplantation of the nucleus*

These experiments demonstrate that a specialised (differentiated) cell has conserved the same DNA as a non-differentiated cell. In this type of study an intestinal cell from a tadpole is chosen as the differentiated cell. The nucleus containing the DNA is extracted and injected into a batrachian egg, the DNA of which has previously been destroyed. A new cell is thus obtained, formed from the nucleus of the intestinal cell and the cytoplasm of the egg. When the cellular division of this new cell is stimulated, a tadpole, and even a perfectly viable adult frog are obtained.

If the injected DNA originating from the specialised cell had only contained the gene of an intestinal cell, cellular division would only have given intestinal cells similar to each other.

In a differentiated cell, the genes have not been lost during cell differentiation since a specialised cell in an intestinal cell possesses all the genes needed to make a complete individual. Thus, all the genes exist but certain ones are repressed.

This experiment, which is relatively simple to perform in bactrachians is (still) totally impossible in humans. One could not, for example, take a skin cell and induce it to replicate to produce an individual the image of the donor (which has not prevented authors in the last few years from imagining this possibility and from giving memorable interviews).

It is important to understand that:

- In a bacterium which is formed from only one cell, all the proteins needed by this bacterium will be synthesised: all the structural genes will be expressed.
- In a higher eukaryote (in a mammal for example) very many cells exist. These cells specialise in their synthesis, and only certain genes are expressed while others are repressed. Evidently, certainly genes repressed in one type of cell will be capable of being expressed in another type of cell and vice versa.

We have seen, in the case of immunoglobulin synthesis, that those genes coding for immunoglobulins are rearranged in the course of differentiation. The sequence in DNA nucleotides is therefore different in an embryonic cell and in a B lymphocyte, but only in the region coding for immunoglobulins. The previous statement, that not all genes are expressed, remains true for all the other genes (about 50 000 in a human cell): a B lymphocyte possesses many non-expressed genes (which would theoretically be capable of coding for many different proteins).

2. *Hypomethylation and gene expression*

We still do not really know what, in a chromosome, facilitates or prevents the expression of a gene. It has seemed, in recent years, that the DNA of mammals can be methylated, and that this methylation probably plays an important role in the expression (or rather the non-expression!) of genes.

(a) Where does methylation take place in eukaryotes?

Only one base is capable of being methylated in eukaryotes: cytosine. This cytosine is most frequently methylated in sequences of type CGCC or CCGG. Thus about 3% of cytosine would be methylated.

(b) Consequences of the methylation of cytosine

This methylation of cytosine in certain of the gene sequences can be considered as a 'bolting in' of gene expression. In fact, if hypomethylation favours expression, then, conversely, methylation prevents the gene from being expressed. Thus, the DNA of genes coding for a protein has been found methylated in an organ which does not synthesise this protein, and not methylated in an organ where synthesis does occur.

These are recent discoveries. It would appear that hypomethylation is a necessary condition, but nevertheless it is not sufficient to determine the expression of genes. Moreover, this is not an absolute rule and several exceptions are known.

Hypotheses have been formulated to try to understand this methylation effect on the blocking of expression. It is possible that methyl prevents fixation of this part of the DNA to proteins needed for the expression of the gene.

(c) The 'DNA maintenance methylases'

A problem is posed with regard to the replication of sequences of methylated DNA. In fact, at the time of replication, the cytosine is always incorporated in non-methylated form. After the initial separation of methylated DNA into its two strands, DNA is formed where one strand contains methylated cytosines, whereas the other strand contains non-methylated cytosines.

Enzymes called 'DNA maintenance methylases' exist, capable of maintaining the information contained in the parent molecule of DNA. These enzymes methylate the cytosines on the new strand, corresponding to the methylcytosines of the other strand.

It is possible that methylation also plays a role in carcinogenesis. Researchers have recently found that the *ras* oncogene obtained from colonic and pulmonary tumours was (in six out of eight cases) hypomethylated. The carcinogens could be inhibitors of DNA methylases, and thus be responsible for an overexpression of oncogenes compared with proto-oncogenes (whose methylation rate is higher than that of oncogenes). Effectively, the 5-azacytidine which produces experimental tumours in rats is an inhibitor of DNA maintenance methylases.

Studies of segments of methylated (or non-methylated) DNA are facilitated by the use of restriction enzymes which recognize (or do not recognise) their palindrome when it is methylated. For example:

Msp 1 recognises CCGG and $C\overset{*}{C}GG$
Hpa II recognises CCGG and not $CC\overset{*}{G}G$

where * denotes a methylated base.

3. DNA sequences controlling the expression of certain genes in higher eukaryotes ('enhancers and silencers')

In recent years the existence of 'enhancers' or activator sequences has been discovered. These are DNA sequences which act on the promoter and stimulate expression. They possess the following characteristics:

— they are active at a distance from the gene (it may be a distance of several thousand nucleotides),
— they can be situated upstream or downstream of the gene,

— they are regulated by diffusable factors.

The 'enhancers' (first described in viruses) have actually been detected in man in the genes for immunoglobulins.

There are also sequences, called 'silencers' or extinguishing sequences, having, as their name indicates, an effect opposed to that of the 'enhancers'. These sequences are also active at a distance from the promoter and can be either upstream or downstream from it.

4. *Comment on thermic shock proteins*

In recent years Jacob (Nobel prize for physiology and medicine 1965) and his team have discovered proteins called 'thermic shock' or HSP (heat shock proteins). These proteins were detected in the mouse and appear after the differentiated cells have been exposed to various chemical or physical stresses (e.g. a thermic shock, from which they get their name). These HSPs then prepare to restart life in the cell. The synthesis of analogous proteins has been shown in the mouse embryo at a very early stage of initial cell division.

C. Hormones

The hormones play an important role in the regulation of protein synthesis in vertebrates, and particularly in man.

1. *Definition*

Hormones are substances produced by an endocrine gland, secreted into the blood, and which travel via the blood to their site of action. A hormone does not act on just any cell, but only on certain cells called 'target cells'.

2. *Mechanism of action*

There are several kinds of hormonal action. We shall only briefly summarise two types.

(a) Hormones which act without entering the cell

These are water-soluble hormones and cannot therefore pass through the cells' lipid membrane. Examples are adrenaline and glucagon.

It is remarkable that this type of hormone can act in cells without penetrating them. The hormone acts through the intermediary cAMP whose formation inside the cell it provokes. The latter acts as a 'relay runner' of hormonal action, and has therefore been termed 'second messenger'.

Having arrived at a target cell, the hormone is recognised by a receptor situated on the outside of the target-cell membrane. The hormone–receptor linkage will initiate reactions culminating in the formation of cAMP. After a cascading series of reactions the cAMP will activate proteins already present in the cell.

cAMP is the best known of the 'second messengers', but other types of 'second messenger' such as IP3 (inositol triphosphate) and DG (diacylglycerol) have been discovered in the last few years.

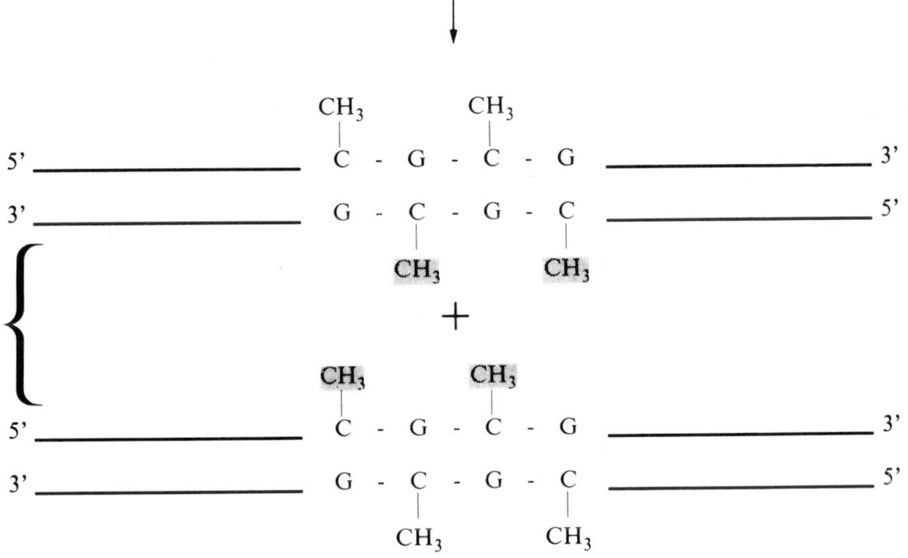

Fig. 89 — Role of 'maintenance DNA methylases'.

(b) Hormones entering the cell

This time we are dealing with hormones, of low molecular weight, soluble in lipids, which can therefore pass across the cell membrane. These are steroidal hormones (for example oestrogens).

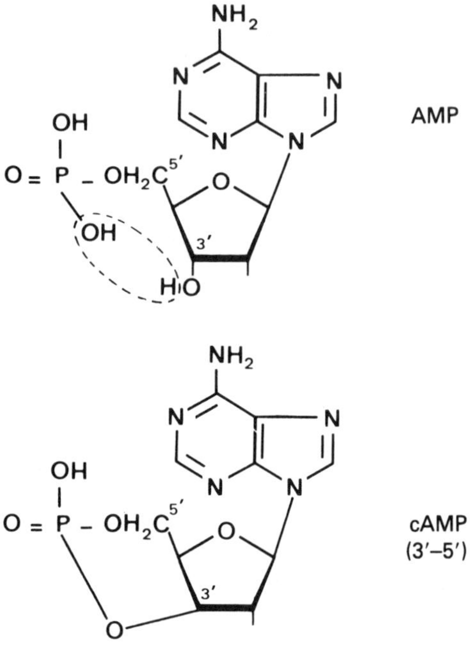

Fig. 90 — Cyclic AMP 3′–5′.

Thus, this type of lipid-soluble hormone, can penetrate into the target cells without difficulty. (Instead the problem, is their transport in the blood which is an aqueous medium. These hormones will be carried in blood by a protein transporter.)

When the hormone has penetrated a target cell, it combines with a specific receptor which only exists in target cells. A target-type cell contains about 10 000 steroid receptors. It must be noted that, unlike the previous case, this receptor is not on the membrane but inside the cell (either in the cytoplasm or possibly even in the nucleus). This hormone–receptor binding provokes a modification of the receptor's shape (an allosteric change). The receptor acquires an affinity with the DNA. The transcription of a gene (or of several genes) is thus initiated and will end in protein synthesis.

Summary

In the first group, that of water-soluble hormones, the hormones:

- remain outside the cell (and will have to act through the mediation of a second messenger, for example cAMP),

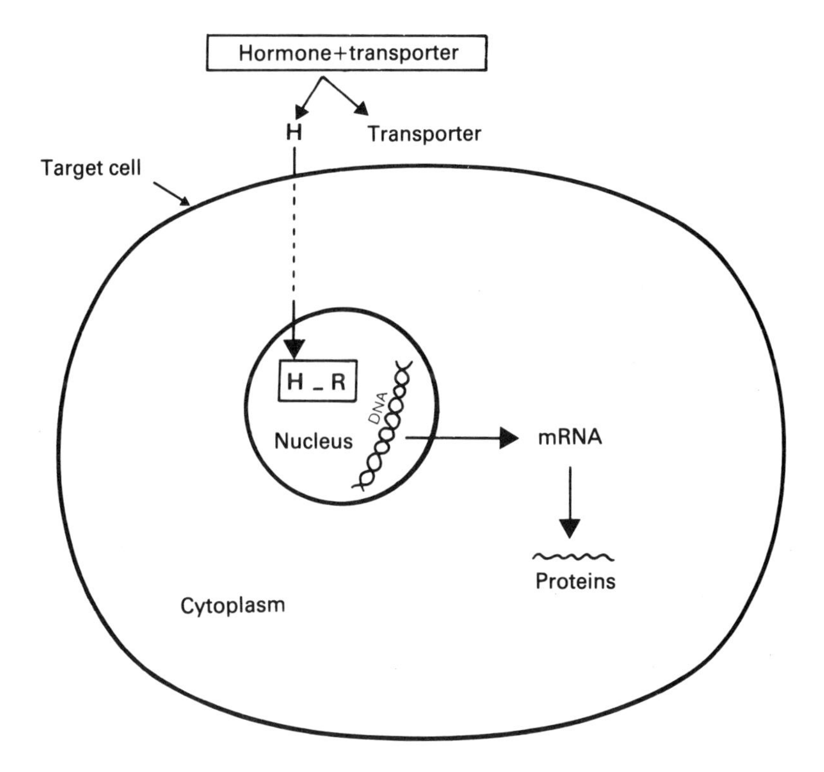

Fig. 91 — Diagram of the mode of action of steroidal hormones (H, hormone; R, receptor).

- initiate *the activation of a protein already present in the cell*

In the second group, that of lipid-soluble hormones, the hormones:

- penetrate into the cell as far as the nucleus,
- initiate *the synthesis of one* or several *protein(s)*.

4

Replication

The information contained in DNA, and required for the synthesis of different proteins, will be transmitted from generation to generation by the process of replication.

When a cell divides to give two daughter cells, the DNA of these daughter cells must be the exact replica of the DNA of the mother cell, whence comes the name 'replication'. It is said that these daughter cells must possess the same genetic patrimony as the mother cells.

Before cell division, or mitosis, the chromosomes begin by splitting into two. In fact this is not a fission in the real sense of the word. It is more precisely a doubling of the DNA; new molecules of DNA are actually created, synthesised. It is this phase of mitosis which particularly interests the biochemist. At each replication the quantity of DNA is doubled.

I. REPLICATION IN PROKARYOTES

A. What is the fundamental characteristic of replication?

Replication is said to be 'semi-conservative': that means that on the two strands of all DNA molecules there is always one strand of old DNA (which comes from one of the two strands of parental DNA).

In fact, at each replication, a separation of the two strands of parental DNA takes place, and simultaneously a copy of each parental strand provides a new strand which remains linked to the parental strand. Two molecules of DNA are thus obtained, each of the two containing a parental strand and a 'daughter'-strand.

B. Elements needed for replication

1. The need for parental DNA

By definition, to create a replica there must be a model. There is a model for the synthesis of DNA. Replication always follows this model, said to be a 'matrix' of DNA. In fact, not only does one strand of DNA serve as the model, but this model is even preserved in the new DNA molecule.

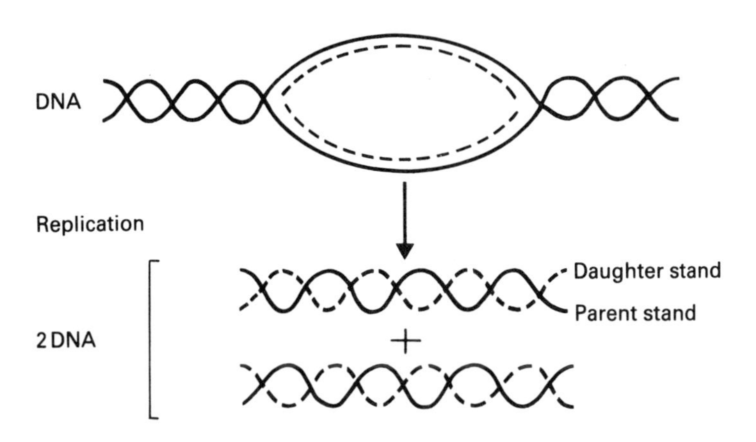

Fig. 92 — Replication is semi-conservative.

This seems simple, but it is not so easy to understand how the strand which serves as the model is preserved. Imagine a goldsmith who, in order to reproduce a twined necklace, separates it into two so as to copy each of the strands separately and thus reform two identical necklaces each having an old and a new strand! Or a potter who breaks a pot in two and forms two pots by completing each old half with a new, complementary half.

2. *The need for nucleotides*

As DNA consists of nucleotides, it is obviously necessary for nucleotides to be present, and, furthermore, for them to be nucleotides specific to DNA:

— deoxyribose,
— with ACG and T (and not U),
— in the form of triphosphate nucleosides, dATP, dTTP, dCTP, dGTP.

These triphosphate nucleosides will then supply the energy needed for the reaction.

3. *The need for enzymes*

At the time of replication, very many enzymes will intervene, for example:

— to allow the two strands of DNA to separate,
— to link the nucleotides to each other,

and also for other reactions which will be discussed later.

4. *The need for bivalent cations*

The presence of Mg^{2+} is indispensable to the synthesis of DNA.

C. Replication mechanisms

1. Synthesis of new nucleotides

Replication takes place:

- in the 5′→3′ direction,
- in complementary fashion, according to the classic rules of pairing: A–T C–G, and
- antiparallel.

2. The propagation of replication is bidirectional

This means that replication is propagated simultaneously to the right and to the left from the point of initiation. (Whereas in transcription the progression of transcription could be compared to an eye which advances along the DNA, here the progress of replication could be compared to an eye which gets larger until replication ceases).

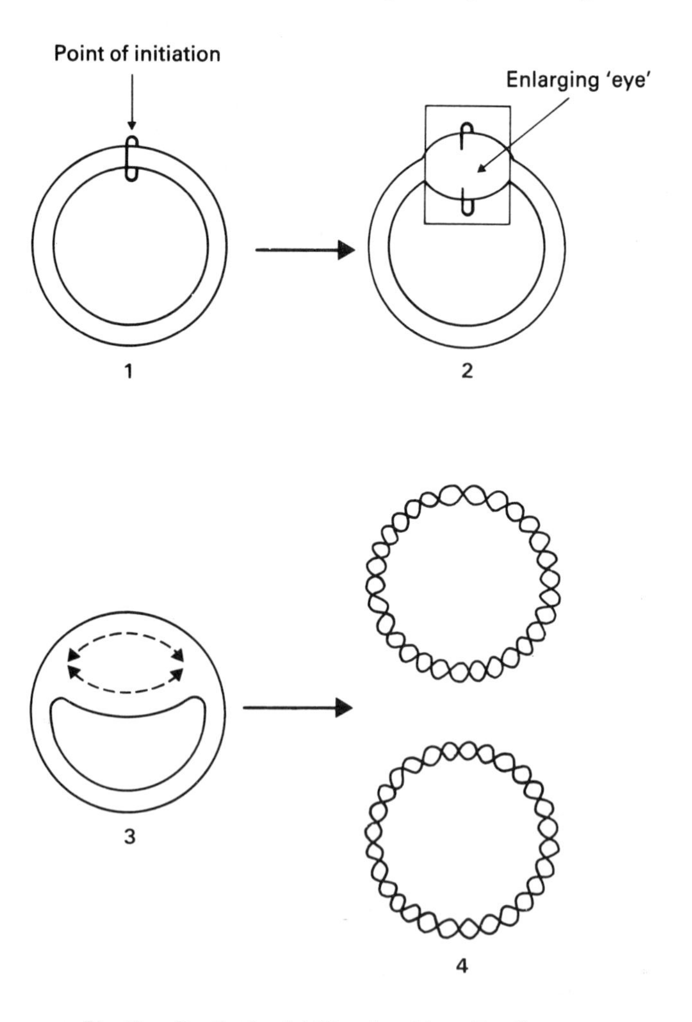

Fig. 93 — Replication is bidirectional (e.g. *E.coli*).

3. Replication is discontinuous for one of the two strands

Replication is continuous for one strand (the strand said to be 'leading') and discontinuous for the other (the strand said to be 'lagging').

In fact, there is a problem for the retarded strand: how can synthesis of this strand proceed from left to right (direction of the propagation of replication in the example below) and at the same time be antiparallel to the parental strand?

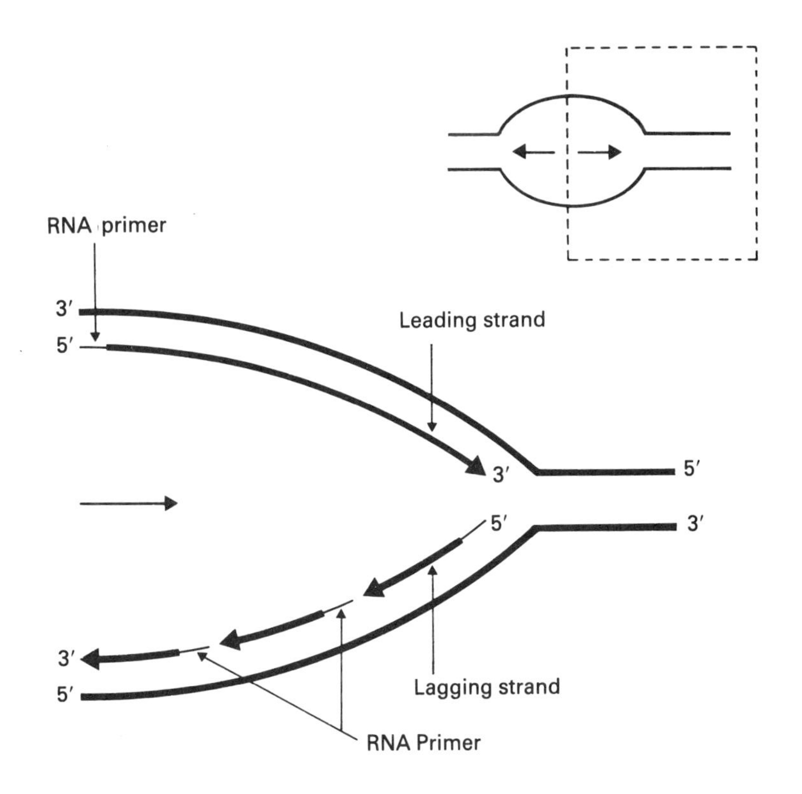

Fig. 94 — Replication is discontinous for one of the two strands.

The solution to this problem was found when it was realised that synthesis of this strand was discontinous. In fact, this synthesis takes place by successive addition in the direction of propagation (from left to right in the example chosen here) of small fragments of DNA. These small fragments of DNA are synthesised 'backwards' ('turning their backs' to the general direction of propagation), but, because of this, are synthesised in an antiparallel way (in relation to the model strand of DNA).

Each small fragment is thus synthesised in the $5' \rightarrow 3'$ direction, but the protraction (discontinuous) of the new strand occurs in the propagation direction.

This discontinuous synthesis is slightly behind the continuous synthesis of the other strand which is how they became known as 'lagging' and 'leading'.

4. *The need for RNA primers*

(a) Intervention of mRNA polymerase (primase)

It could be supposed that the piling up of nucleotides on each other to form a strand of DNA would occur through the action of an enzyme, a DNA polymerase. This is partly so, but there is a problem nevertheless: the DNA polymerase has no 'initiative'; it does not know how to initiate a chain. It is only able to elongate a chain of nucleotides, i.e. it is only capable of adding a nucleotide to the 3′OH end of a nucleic acid. It matters little whether this nucleic acid is DNA or RNA. So how does synthesis of a strand of DNA begin?

This is where the RNA polymerase intervenes. In fact the RNA polymerase is itself capable of starting a nucleic acid chain (but clearly an RNA chain, not DNA).

Synthesis of a new strand of DNA therefore begins with a small fragment of RNA — called an RNA 'primer' — produced by an RNA polymerase (called RNA primase). The DNA polymerase (III) will then take over, lengthening the primer, but this time with DNA.

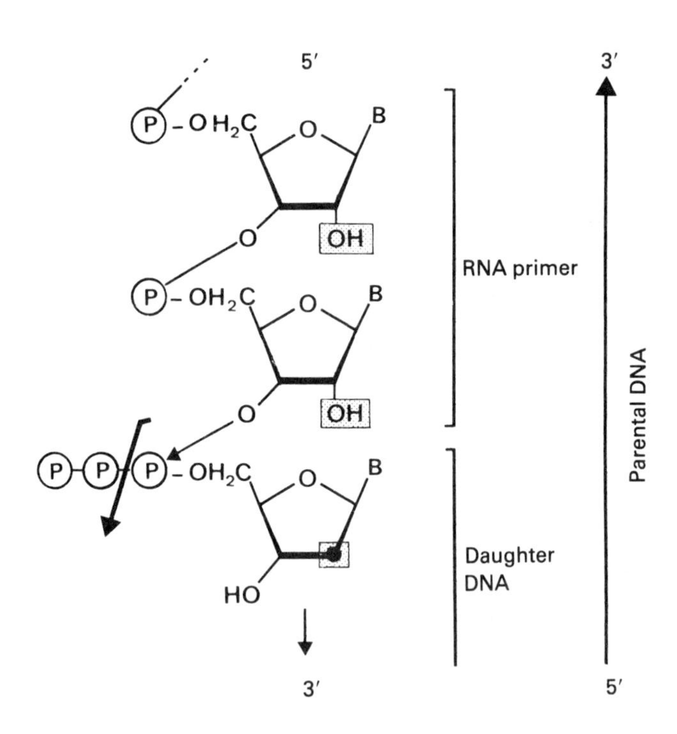

Fig. 95 — Need for RNA primers to initiate replication.

(b) Hydrolysis and replacement of RNA primers

RNA primers are next destroyed by hydrolysis and replaced by DNA. This is done by a certain enzyme which both removes the RNA primer and resynthesises DNA in their place. It is a DNA polymerase (I) endowed with 5′→3′ exonucleasic properties (which serve to hydrolyse the RNA primers) and also with polymerasic properties (allowing deoxyribonucleotides to be added at the 3′ end of the previous DNA fragment). This DNA polymerase may be thought of as a small beaver which advances, gnawing the RNA primer from the fragment in front of him, while building with his hind paws a chain of nucleotides, lengthening the fragment of DNA behind him until the two segments of DNA are adjacent.

Finally, the fragments of DNA, liberated from their RNA primer and now contiguous, will be fused to each other by a ligase enzyme.

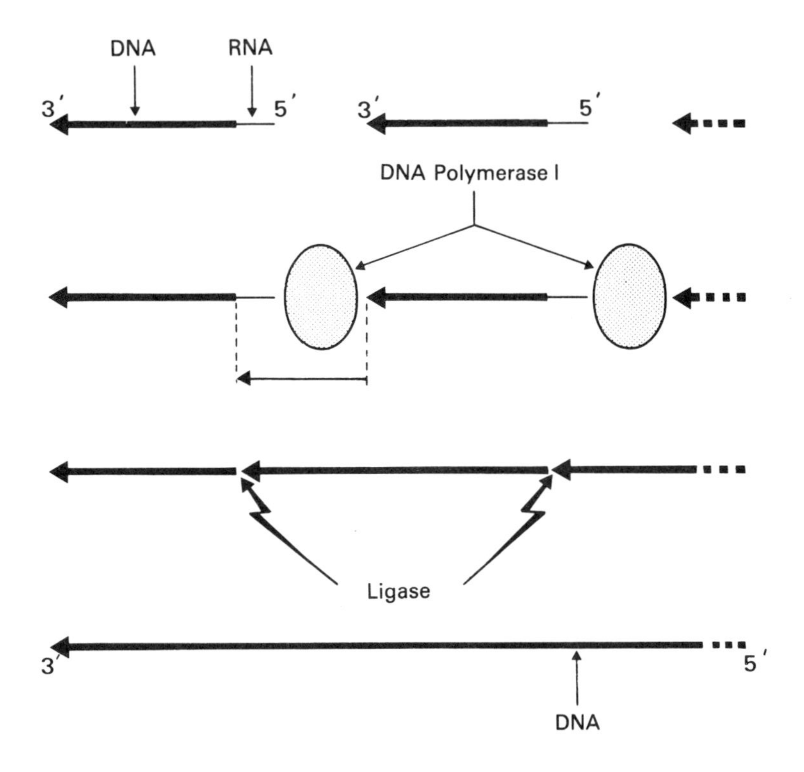

Fig. 96 — Hydrolysis and replacement of RNA primers.

5. Comment on editing enzymes

It is remarkable that an enzyme (such as DNA polymerase I) should be endowed with both (exo)nucleasic and polymerasic properties. DNA polymerase III also possesses such properties.

DNA polymerases III and I can also act as 3′→5′ exonucleases (in addition to their 5′→3′ exonuclease action). They are in fact endowed with an 'editor function' which can 're-read' the last nucleotide set in place. If by chance there is a fault in pairing, they remove this last nucleotide (using their exonuclease 3′→5′ properties) and append the appropriate nucleotide (using their polymerasic properties). You now see why DNA polymerases cannot initiate a DNA chain: they must necessarily be able to re-read the last nucleotide laid down and verify that this nucleotide is well matched according to the rule of complementarity to the nucleotide of the antiparallel strand.

This mechanism ensures a very faithful replication. It is not the first time that we have discussed enzymes which can edit. We also saw this astonishing property with aa~tRNA synthetases. However, enzymes with the editing function are expensive (it is for this reason that the RNA polymerases do not possess such properties. An error during the synthesis of a molecule of mRNA does not have such serious consequences as an error during replication). There is always a compromise between maintaining life and conserving energy.

The way in which the helix of the parental DNA molecule separates into its two strands is complex and still not well understood. Many enzymes intervene, in particular the gyrase which absorbs the positive supertwists created by the opening of the double helix. We have already seen (during transcription) that the gyrase is capable of creating negative supertwists beforehand which will then allow compensation for the positive supertwists produced by the unrolling of the helix.

Proteins such as the SSB (single-strand DNA binding) proteins also intervene, (called 'helix destabilising proteins'). These proteins attach themselves to each of the chains of the parental helix as soon as separation begins. They thus prevent the two chains from re-pairing; moreover, they also prevent a chain from folding back on itself, to form a loop, which would occur in self-complementary regions.

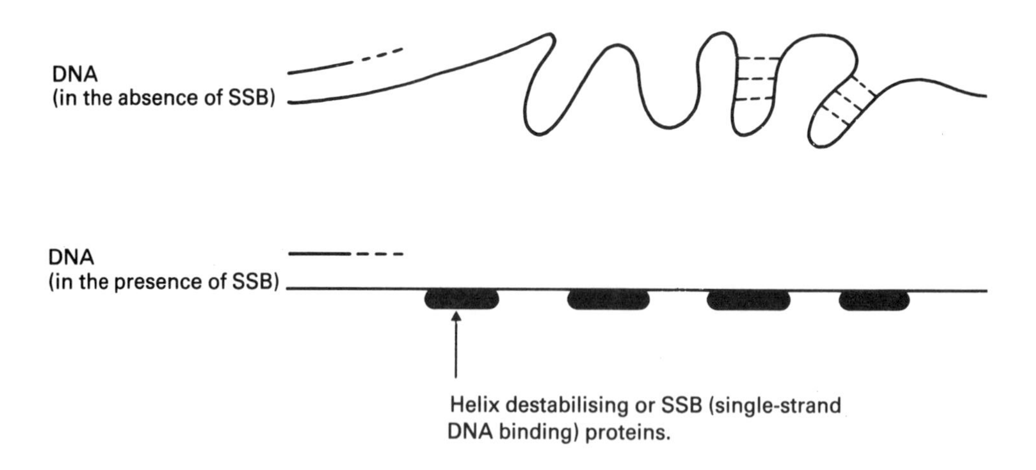

Fig. 97 — Helix destabilising proteins.

To sum up, the following are involved in replication:

— gyrase and SSB proteins: which facilitate the separation of the double helix,
— primase: which allows synthesis of RNA primers,

— DNA polymerase III: which synthesises DNA on the primer,
— DNA polymerase I: which hydrolyses the primers and replaces them with DNA
— ligase: which reunites the DNA fragments.

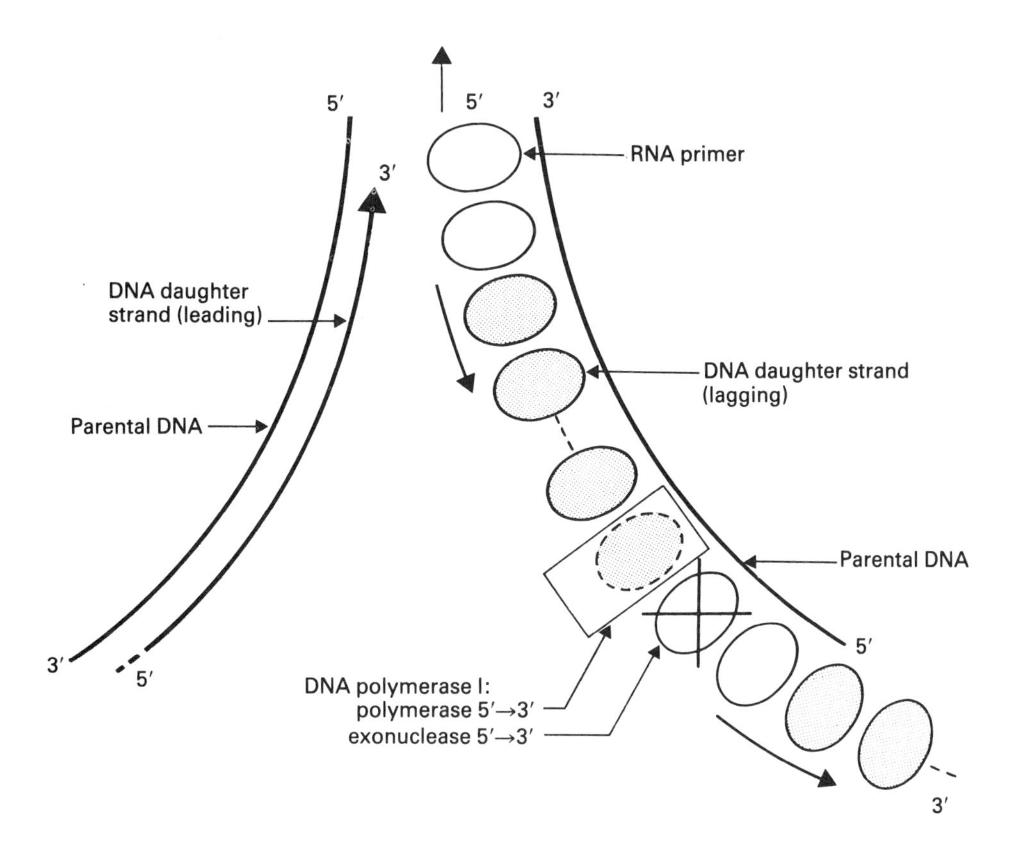

Fig. 98 — Diagram summing up the role of the principal enzymes intervening in replication of DNA.

II. REPLICATION IN EUKARYOTES

The mechanism of replication in eukaryotes is comparable to that of prokaryotes. It takes place:

— bidirectionally,
— in a complementary, antiparallel way in the 5′→3′ direction,
— discontinuously for one of the two strands,
— with RNA primers.

Rather than having only one initiation point as was the case with prokaryotes, replication in eukaryotes starts simultaneously at several points on the same chromosome. Thus there are several thousand initiation points. This is rendered necessary because of the great length of eukaryote DNA.

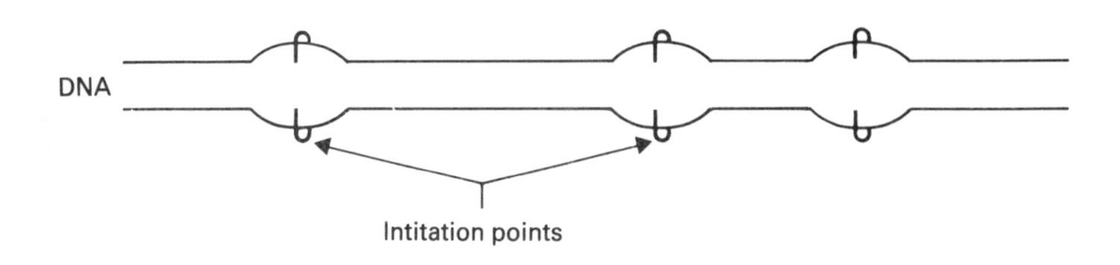

Fig. 99 — Replication in eukaryotes.

The different stages of replication in eukaryotes are still not well understood. The study of replication is difficult because of the association of DNA with the histones which must be simultaneously synthesised and divided between the new and old strands. Moreover the separation of the helix is also very complex. Remember that the gyrase has not been identified in eukaryotes.

III. MUTATIONS

A Definition

Mutations are copying accidents in the bases, which occur during the replication of DNA. Newly synthesised DNA will therefore no longer be an exact replica of the parental DNA. The accident in copying is not merely a badly copied base, but can also be a forgotten base or, conversely, an additional inserted base. It is entirely comparable to 'misprints' which occur at the printer's where a letter can be replaced by another, omitted, or added. Mutations are also known to appear where blocks of sequence are rearranged and intervene in chains of DNA already synthesised.

If the mutation takes place in a gene, the mRNA transcribed from this gene will be similarly modified. Depending on the case, the synthesised protein may or may not be very different from the protein initially coded by the non-mutated gene.

Errors in copying can also be produced during transcription, but there it is not a serious problem because each mRNA has a short life and if a mRNA has undergone mutation, it is negligible in the face of the thousands of mRNAs which have not.

There are different kinds of DNA mutation. It must be understood that any mutation on a part of the DNA which is expressed automatically brings about a change in the codon. It is for this reason that the different possible mutations will be considered logically at codon level, and so in the mRNA (although the origin of the error may be a mutation at DNA level).

B. Different types of mutation

1. Mutations without change in the reading frame

(a) Mutations with no effect, called 'silent'
A codon is replaced by a codon coding for the same amino acid. For example, UUU is replaced by UUC. These two codons code for Phe. This mutation therefore has no effect.

(b) Conservative mutations
A codon coding for an amino acid is replaced by a codon giving an amino acid of the same group. For example AAA (Lys) is mutated into AGA (Arg). Lys and Arg form part of the same group of amino acids (basic amino acids) and this mutation is usually without effect.

(c) False direction mutations
A codon is replaced by a codon giving an amino acid which is chemically very different. For example AAG (Lys) is changed to GAG (Glu). Lys is a basic amino acid whereas Glu is an acidic amino acid. Usually an abnormal protein results.

(d) Mutation affecting the stop codon
The mutation transforms the codon coding for an amino acid into a stop codon. For example, UGC which codes for Cys is changed into UGS which is a stop codon. If the error occurs at the beginning of the peptide chain, the consequences are clearly serious. However, if the error occurs at the end of the chain it could be negligible.

Conversely, a stop codon may be transformed into a codon coding for an amino acid. A longer protein will then result.

2. Mutations with a change in the reading frame

A mutation with a change in the reading frame, through deletion or insertion of a base, produces a shift in the recognition of triplets. These mutations are serious if the shift occurs at the start. In this case a completely different protein is actually obtained.

An example of shifting:
Suppose the example relates to the sequence:

AUGGCCUCUAACCAUGGCAUA

1. Reading according to the first frame

AUG	GCC	UCU	AAC	CAU	GGC	AUA
Met	Ala	Ser	Asn	His	Gly	Ile

2. Reading according to the end frame (after deletion of a base — e.g. G in fourth position): a shift of one

AUG	CCU	CUA	ACC	AUG	GCA	UA
Met	Pro	Leu	Thr	Met	Ala	

3. Reading according to the third frame (after a new deletion — e.g. C formerly in fifth position): a shift of two

AUG	CUC	UAA	CCA	UGG	CAU	A
Met	Leu	stop				

4. Reading according to the initial reading frame (after a new deletion — e.g. C formerly in the sixth position): a shift of three, i.e. a return to the initial reading frame (only an amino acid is suppressed, the rest of the sequence is identical to that obtained initially).

AUG	UCU	AAC	CAU	GGC	AUA
Met	Ser	Asn	His	Gly	Ile

C. Some consequences of mutations

1. Diseases

The mutation of a single base, taking place in a structural gene, can bring about serious pathological disorders. A classic example is that of a disease (sickle cell anaemia) caused by an abnormality of one of the peptide chains of haemoglobin. The replacement of a single amino acid on this peptide chain upsets the properties of haemoglobin, which becomes incapable of correctly playing its role of liberating oxygen to the tissues. This disease will be discussed further in the chapter on genetic engineering.

DNA of a normal cell: 5′. . .GAG. . .3′
3′. . .CTC. . .5′ (transcribed strand)
(code for Glu)

Mutated DNA (sickle cell anaemia): 5′. . .GTG. . .3′
3′. . .CAC. . .5′ (transcribed strand)
(code for Val).

Remember also that with a little practice it would be possible to reason from the sequence of the untranscribed DNA strand (which is identical as to polarity and bases — with the exception of U replaced by T — to mRNA).

What of hereditary transmission; is a mutation on the DNA transmitted to descendants? The answer is very different depending on whether prokaryotes or eukaryotes are involved. In *E.coli*, for example, the mutation is transmitted automatically to descendants. In the higher eukaryotes mutation is only transmitted to descendants if the mutation has affected the sexual cells.

2. *Theory of evolution*

A mutation is not always unfortunate. It is moreover one of the foundations of the 'theory of evolution' that mutations occur randomly. If the mutation represents an advantage (and if, indeed, this mutation occurs in sexual cells) the mutated individuals and their descendants will survive better than non-mutated individuals.

D. Mutagenic agents

Mutations occurring spontaneously at the time of replication are rare. They are due to an error on the pairing of bases. The high degree of fidelity of replication is due to the fact that DNA polymerases I and II are endowed, as previously discussed, with the 'editor' function (reading of proofs and correction of errors).

However, agents called 'mutagens' exist which increase the number of mutations.

1. *Chemical agents*

— Chemical substances which transform the base, for example NH_2 into OH, thus giving uracil.
— Chemical substances which resemble a base; these substances will be introduced into DNA and during replication the substances will be 'copied'. A supplementary base will then be obtained, which will provoke a mutation by insertion.

2. *Physical agents*

Physical agents inducing mutations include radiations such as X-rays or ultra-violet rays.

Excessive exposure to the sun is dangerous, and can be responsible for cancers of the skin. The energy of ultra-violet rays is mainly absorbed by the pyrimidines of DNA. Phenomena of electron excitation result, culminating in a modification of the type of bond.

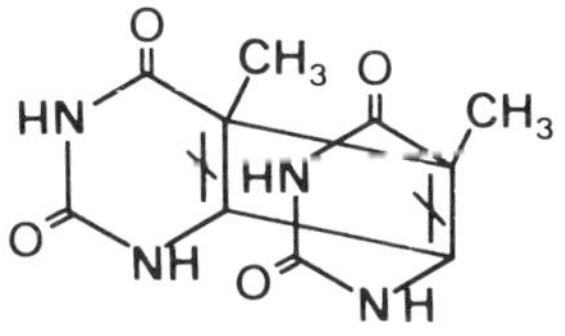

Fig. 100 — Thymine dimer.

Thus, where two adjacent pyrimidines (in particular two thymines, TT) exist on one strand of DNA, these two bases, under the influence of ultra-violet rays, will bond to each other covalently. Instead of two adjacent thymines being matched by hydrogen bonds to complementary bases (AA) situated on the complementary strand, they form a dimer. The thymine dimers bring about a local distortion on the DNA. The DNA polymerase will not know how to copy these two thymines and replication will therefore be blocked at this level.

E. Repair of thymine dimers

1. Repair by excision–resynthesis

'Repair enzymes' exist, capable of removing the defective area (by cutting some nucleotides before and some nucleotides after this defective zone). Other enzymes must then intervene to repair the gap. This repair of the DNA is affected in *E.coli* by DNA polymerase I. This enzyme extends the interrupted DNA across the gap, proceeding to a complementary copy from the information contained in the intact strand.

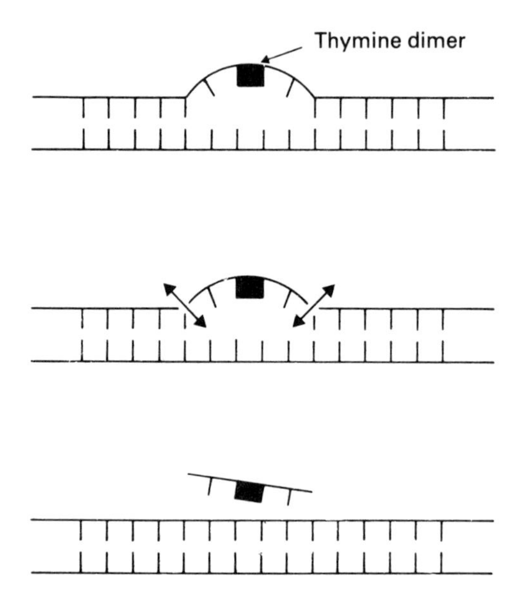

Fig. 101 — DNA repair by excision–resynthesis.

A ligase must then intervene to fuse two fragments of DNA.

The system of repair exists not only in prokaryotes but also in man. There is serious disease called 'xeroderma pigmentosa', happily very rare, which is due to a defect of excision–resynthesis enzymes.

One cannot but help admiring the safety with which prokaryotes and eukaryotes maintain intact the information from the DNA. A molecule of DNA with two complementary strands in fact contains double information. If a strand is damaged, resynthesis of this strand will always be possible. Thanks to the second intact strand the information is not lost.

mRNA, with a single strand, does not have this advantage. (However, as discussed above, what is one damaged mRNA beside thousands of intact mRNAs?) In the same way, viruses with single-stranded DNA do not know the security of cells which possess a double-stranded DNA.

Excision enzymes are synthesised in sufficient quantity to effect the necessary repairs. However, if the lesions are too numerous, this system is inadequate. Other types of repair exist, but they are at present only known in *E.coli.*

2. Post-replicative repairs

(a) The post-replicative gap

If the system of repair with excision–resynthesis enzymes is over-run the DNA polymerases will skip the lesion, thus producing a breach called a 'post-replicative gap'. A critical situation results:

— the parental strand contains a thymine dimer
— the daughter strand contains a gap. The information (on a short fragment of DNA) is thus totally lost to each of the two strands. Fortunately the second parental strand and its son-strand are close by and intact.

(b) The recA protein

• Mode of action

The recA protein ('rec' for recombination) is a protein which plays an important role in allowing a recombination or, in this case, an exchange, between two homologous segments of DNA (the second parental strand and the first daughter strand).

This is not a repair in the strict sense of the word, once this recombination now produces a new breach on the second parental strand of DNA. However, this mechanism is interesting none the less.

In the initial situation there is:

— 0 information out of 2 on one side.
— 2 information out of 2 on the other.

After the intervention of the recA protein there is now:

— 1 information out of 2 on one side (information lost because of the thymine dimer)
1 information out of 2 on the other (information lost because of a breach).

There must then be intervention by the excision–resynthesis enzymes, which will

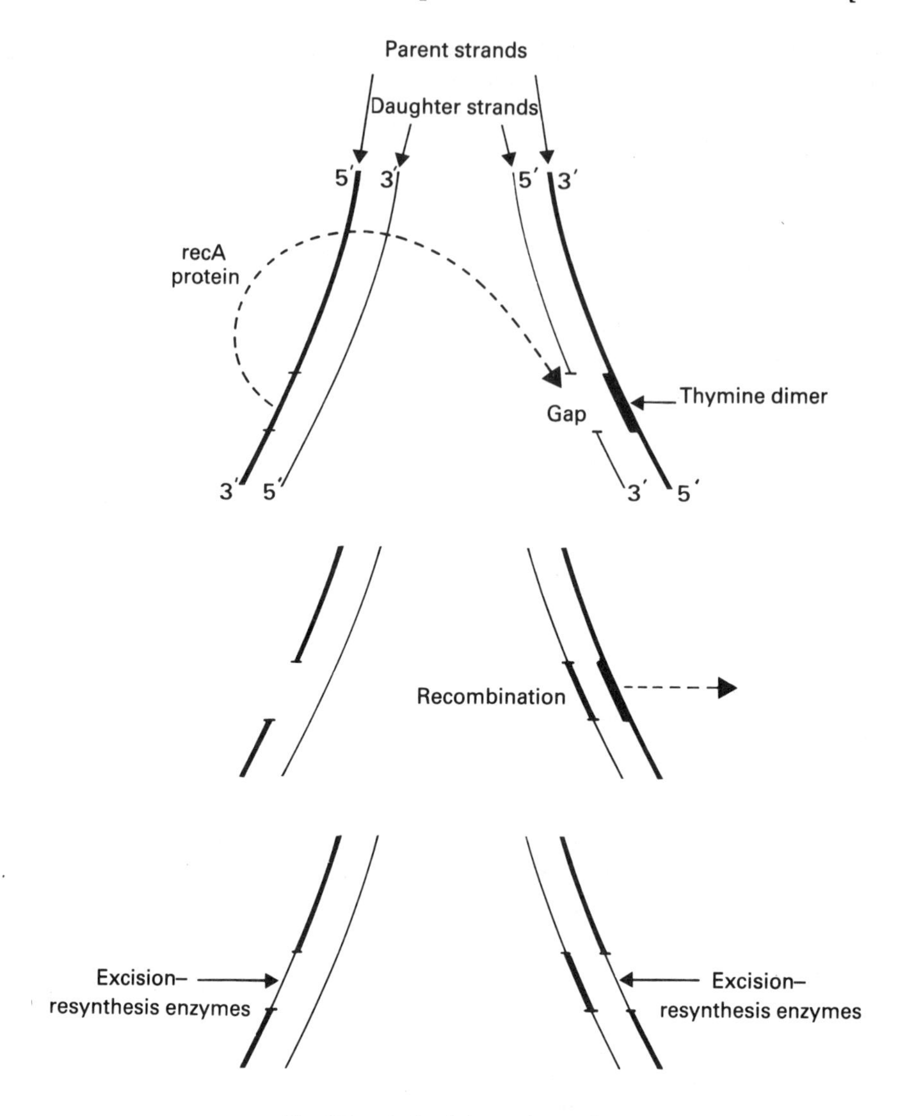

Fig. 102 — Role of the recA protein.

eliminate the thymine dimer on one strand and make good the breach on the other. Each time a copy can be made from information contained on the opposite strand. Finally the ligases will join the divide DNA fragments.

- Repression of the recA synthesis by the recA protein

In a bacterium there are only a few molecules of recA protein, but these are sufficient to effect the necessary recombinations under normal conditions. The number of recA molecules is limited because its synthesis is normally repressed by a repressor called 'lexA'.

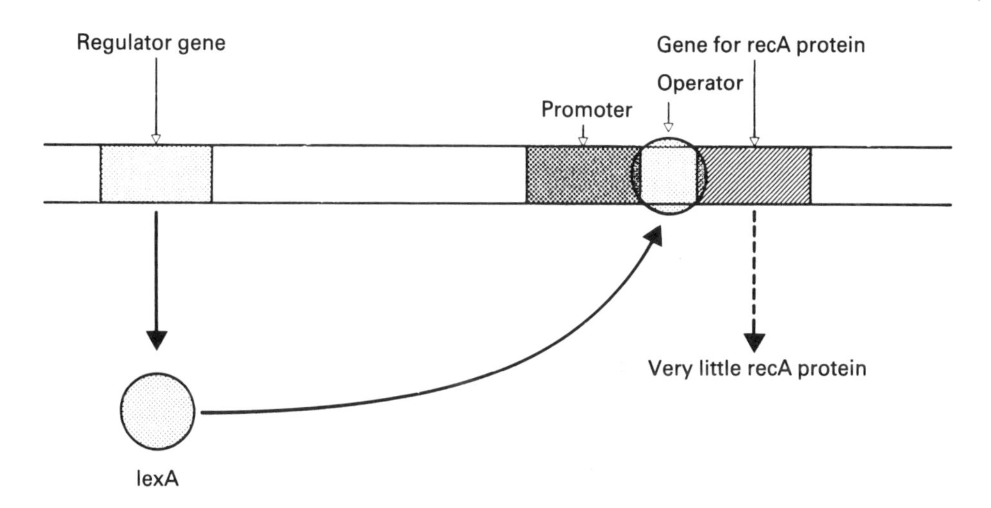

Fig. 103 — Repression of recA protein synthesis by the lexA protein.

3. *The SOS system*

The SOS system is activated when the previous system is overloaded. In the 1970s Radman called this system 'SOS' because it permits cell survival. An arrest of replication would result in cell death. However, we shall see that, if synthesis follows, it is often at the cost of replication errors.

(a) *Proteolysis of lexA by recA*

The recA protein is a remarkable protein which possesses protoelytic properties in addition to the recombinative properties described above. However, the former are only manifested when the recA protein is activated. Activation of the recA protein can be provoked by a thymine dimer or a single-stranded DNA. When the first two systems are overloaded, arrest of replication occurs with the presence of thymine dimers and single-stranded DNA. This acts as a distress signal which activates the proteolytic properties of the recA protein. The recA protein then hydrolyses its repressor, the lexA protein. The number of molecules of recA protein immediately increases which enables the necessary recombinations to take place.

Thus, the recA protein is an inducible protein, whereas the excision–splicing enzymes are constitutive enzymes. The recA protein is not, however, the only protein induced in this case. In fact *E.coli* possesses many inducible SOS genes whose induction is seen when the lexA repressor is proteolysed.

(b) *Replicative errors*

However, the fidelity of replication will be diminished by the induction of the SOS response. In effect, the SOS response occurs very quickly. Trying to be speedy, the cell repairs its DNA badly. Thus, reparative synthesis, confronted by a lesion, takes place by random insertion of non-complementary bases, no longer according to the rules of pairing, and without 'proof reading'. Replication therefore becomes error

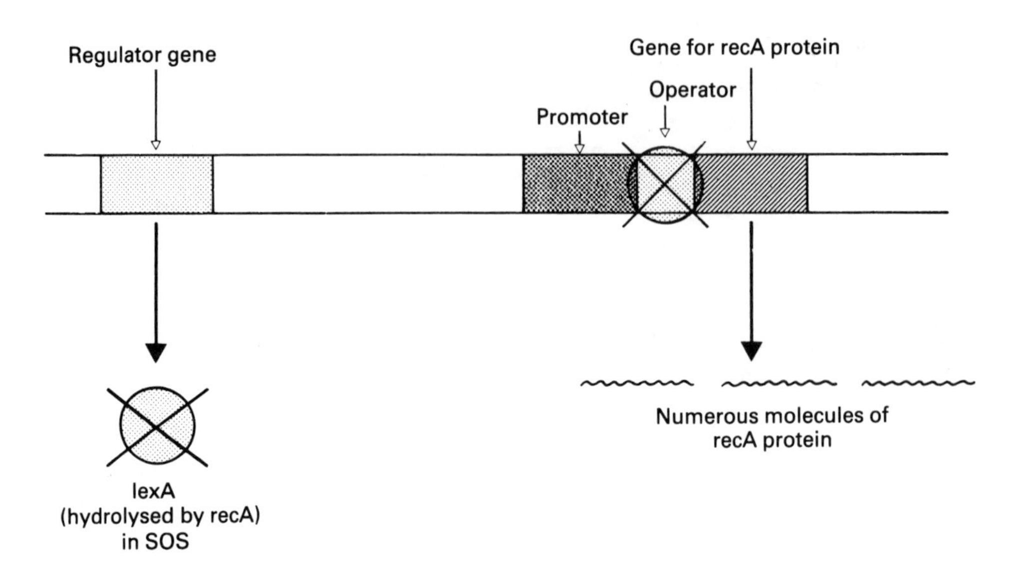

Fig. 104 — The SOS system.

prone. The SOS repair avoids blocking of DNA synthesis and thereby prevents the cell from dying, but this leads to mutations and will be the price to pay for survival . . . .

As already indicated, the two last systems (post-replicative repair and SOS) have not been described in man. However, nature is a great re-exploiter of strategies! It would not be the first time that a basic process, first observed in *E.coli*, is subsequently detected in man.

5

Viruses

Viruses are infectious agents even smaller than bacteria.

I. STRUCTURE

Viruses have a very simple structure. They consist of particles which are called 'virions', comprising

— a simple piece of nucleic acid. The nucleic acid of viruses is either DNA or RNA. Thus, it is possible to distinguish:
- DNA viruses, e.g. hepatitis B virus,
- RNA viruses, e.g. hepatitis A virus, poliomyelitis, influenza or AIDS (acquired immunodeficiency syndrome)

— a proteinous coat or capsid (from the Greek 'capsa' meaning 'box'). This proteinous coat is found on the outside, surrounding and protecting the central nucleic acid against enzymatic attack. In certain small viruses, capsids are not found. However, proteins are found in contact with the viral nucleic acid (RNA), forming what is called the nucleocapsid (also sometimes termed the central nucleoid) or 'core'

— sometimes an envelope. In fact, for certain viruses (such as the hepatitis B virus, retroviruses, etc.), the capsid is itself protected by another protein envelope. One part of the envelope constituents is in fact 'borrowed' from the membrane of the host cell.

II. PATHOGENIC POWER OF VIRUSES

As mentioned above, a virus, is generally just a nucleic acid protected by a protein coat. In a way it is simply a small piece of chromosome. A viral enzyme is sometimes associated with this nucleic acid (like the reverse transcriptase in retroviruses). However, viruses do not possess the equipment (in particular they have no ribosomes) to be able to effect protein syntheses. In order to live and to reproduce, viruses are forced to be parasites, in order to utilise the synthesis systems of the host cell which receives them.

Viruses which parasitise bacteria (called 'bacteriophages') are distinct from viruses which parasitise multicellular organisms (plants, animals and man). Viruses are specific, but the degree of specificity depends on the viral species. Thus, a virus can have an affinity for a host animal and even for particular cells of this host. Cellular receptors and viral proteins are implicated in this recognition. Remember that viruses are responsible in man for many diseases such as measles, mumps, chickenpox, shingles, etc. (besides hepatitis, poliomyelitis, influenza, AIDS, etc., already mentioned above.) The first viruses studied were bacteriophages. In the chapter on the techniques of DNA sequencing, we saw the first sequence known to be viral DNA was, in 1976, that of a bacteriophage (ϕX174).

Great progess has now been made in the knowledge of more complex viruses such as the AIDS virus or the hepatitis B virus.

III. EXAMPLES OF VIRUSES

A. The AIDS virus

1. AIDS

AIDS (acquired immunodeficiency syndrome) is a disease, first described in 1981, which first particularly afflicted homosexuals and people, such as haemophiliacs, receiving many blood transfusions. It is caused by the destruction of certain T lymphocytes (T4 lymphocytes) by a virus. We saw in the section on immunoglobulins that B lymphocytes, through antibodies which they are capable of synthesising, are responsible for humoral immunity. The T lymphocytes do not secrete antibodies but participate in immune defence at the cellular level.

2. The AIDS virus is a retrovirus

The virus responsible for AIDS is a retrovirus, identified by a French group (led by L. Montagnier) as the LAV (lymphadenopathy associated virus), then by an American group (led by R. C. Gallo) as the HTLV-III (human T-lymphotrophic virus III), and also known under the name of ARV (AIDS-related virus). It should now be called 'HIV' (human immunodeficiency virus).

(a) Recap of the structure of retroviruses (reverse transcriptase genes, LTR sequences)

Retroviruses are RNA viruses formed from two identical molecules of RNA arranged top-to-bottom. For simplicity we will only consider one RNA strand and not two. For replication, RNA must pass through an intermediate form to DNA and integrate itself in the DNA of a host cell. This transition of single-stranded RNA to double-stranded DNA is accomplished (by a series of complicated reactions) by a viral enzyme, called reverse transcriptase, discovered by Temin and Baltimore. (They received the Nobel prize for physiology and medicine in 1975.) There is a problem. How can reverse transcriptase synthesise a DNA from an RNA model, when, like all DNA polymerases, it does not know how to initiate DNA synthesis? In fact, we have seen that the DNA polymerase which facilitates DNA replication in prokaryote and eukaryote cells needs a RNA bait. In retroviruses the bait is a tRNA of cellular orgin (present inside virions).

Let us define a little the structure of a retrovirus, before and after its integration into the DNA of a host cell. We may distinguish integration of a retrovirus into the DNA of a host cell:

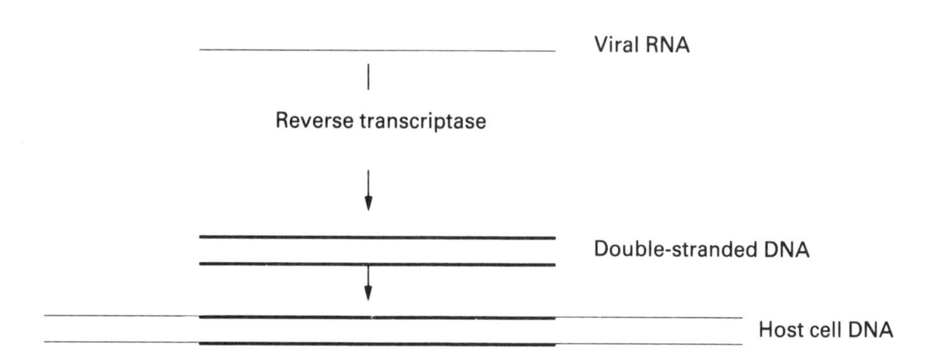

and the molecular structure of the genome of a retrovirus during the different phases of integration:

- Genomic RNA

The RNA of a retrovirus is a single-strand of RNA in which the coding regions are flanked by essential sequences for viral replication and expression. These sequences are, from end 5′ towards 3′:

— R (short for 'repeat'): a short sequence at the 5′ end (the same nucleotides will be repeated at the 3′ end);
— U_5: unique sequence of the 5′ end (consisting of about 80 nucleotides);
— UU: two uracil nucleotides;
— TBS ('tRNA binding site'). This is the site where the tRNA which will use baits during replication, is united (at the 3′ end). The complementary and antiparallel linkages will involve about 20 base pairs.
— NC: a non-coding sequence;
— coding region: this is situated in the centre and contains very few genes. These are:

'*gag*' (for 'group-specific antigen') or the antigen gene of the group which codes for a polyprotein which, when cut up, gives nucleoid proteins. Amongst these internal proteins is carried group antigenicity (for example, murine, avian and feline viruses)

'*pol*' (for 'polymerase') which codes for reverse transcriptase, and

'*env*' (for 'envelope') which codes for the envelope's glycoproteins. One of them carries the antigenicity of each viral serotype.

In the coding region there is also one gene coding for a viral protein (or a protein segment) specific to the retrovirus. (For example in oncogenic retrovir-

- Genomic RNA

5′ R–U_5-UU-TBS-NC-coding region-NC-Pur-AA-U_3-R 3′
(RNA)

- Viral DNA after the action of reverse transcriptase: non-intergrated viral DNA.

5′ AA-U_3-R-U_5-TT-TBS-NC-coding region-NC-Pur-AA-U_3-R-U_5-TT 3′
(DNA)
3′ TT-U_3-R-U_5-AA-TBS-NC-coding region-NC-Pyr-TT-U_3-R-U_5-AA 5′

- Intergrated viral DNA (provirus)

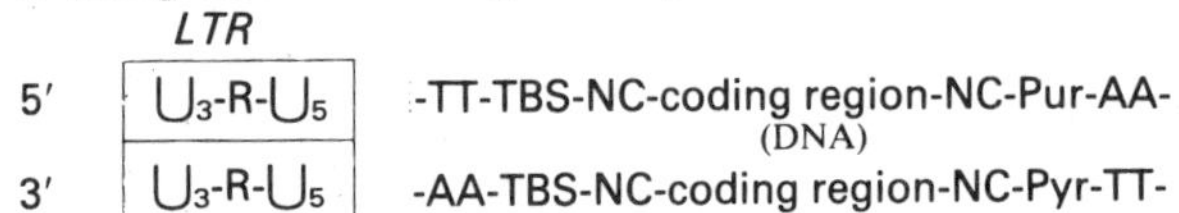

- Host-provirus junction, e.g. AMLV virus

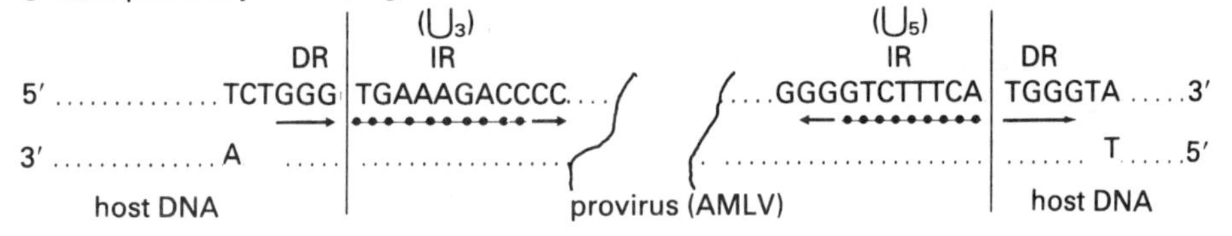

Fig. 105 — Integration of retroviruses into the DNA of a host cell and the molecular structure of their genome (see abbreviations in the text).

uses one gene is found. This is not the case with the AIDS virus which is not an oncogenic retrovirus; it does not cancerise the cells but destroys them).

— NC: a non-coding region;
— Pur: a sequence rich in purine bases;
— AA: two adenine nucleotides;
— U_3: sequence unique to end 3′ (about 300 nucleotides)
— R (short for ‘repeat’) at end 3′

• Viral DNA after the action of reverse transcriptase: 'non-integrated DNA'
After the action of reverse transcriptase the transcribed double-stranded DNA molecule is longer than the corresponding genomic DNA. In fact, there are additions to the two ends of both sequences:

at end 5′ a U_3 sequence is added;
at end 3′ a U_5 sequence is added.

The entire U_5–R–U_5 is called 'LTR' (long terminal repeat): there is therefore one LTR at each of the two ends.

Integrated viral DNA (provirus)
Double-stranded linear DNA becomes a closed circle, then it is re-opened to be integrated into the host-cell's DNA. Integrated viral DNA is called 'provirus'. On integration, the dinucleotides (TT, AA) situated at the ends of transcribed DNA (and forming part of U_3 or U_5) are eliminated.

The junction at the integration site implies:

— in the host DNA, there is a direct repeated sequence, DR ('direct repeat'), at the junction with the provirus.
All the DRs are four to six b.p. in length but their sequences are different. For the same provirus, moreover, different DRs are found in the DNA of a host cell. The integration site of the virus is therefore not specific and the virus can be integrated into the genome of the host at several different places.
— in the viral DNA, there is an inverse repeated sequence, IR ('inverted repeat'), at the extremities of the retrovirus. Two sequences are said to be 'inverse repeated' where one is inverse and at the same time complementary to the other; Fig. 105 shows an example of the host DNA–provirus DNA junction for AMLV (Abelson murine leukaemia virus).

(b) The genes gag, pol, env, and the ORFs of the AIDS virus
To clone, and then sequence an RNA retrovirus, the double-stranded DNA copy (natural or laboratory) formed by the action of reverse transcriptase must be used (see the chapter on genetic engineering). It is this DNA copy which the French team of L. Montagnier, in association with that of P. Tiollais, was able to sequence recently. This was also achieved by three American teams working independently. The French team, in January 1985, published the sequence of this DNA molecule which comprises 9193 nucleotides. The three genes, *gag, pol, env*, are identified here. In addition , two ORFs (open reading frames), i.e. two regions with no stop codon, able as a result, to code for two viral proteins, were identified. These are ORF Q situated between *pol* and *env,* and ORF F. A slight overlap is seen between *gag* and *pol, pol* and Q, *env* and F. The role of proteins coded by the ORFs Q and F is still not understood.

The American team of R. C. Gallo also published in January 1985 (some days later) an article on the sequence of the HTLV-III virus. The sequence was the same as that of the LAV virus (The sequence of ARV virus is very similar but some diffrences are found with the *env* gene).

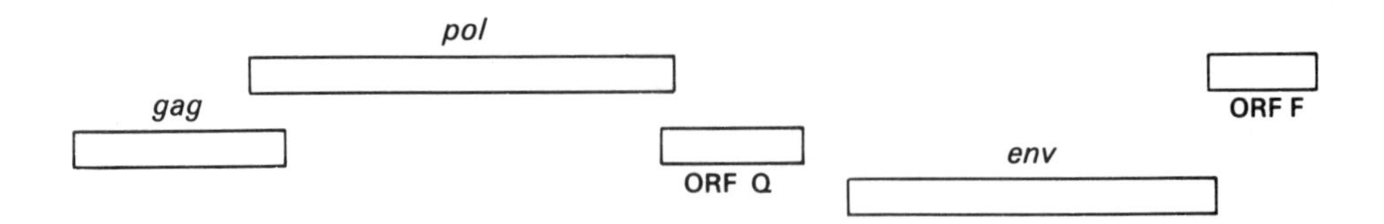

Fig. 106 — Genes of the AIDS virus.

To date, no means of combating AIDS has been discovered. A drug ('HPA 23' ammonium antimony tungstate) is presently being tested. It could act *in vivo* by inhibiting the reverse transcriptase of the virus and could thereby prevent it from multiplying.

B. The hepatitis B virus

1. The original structure of the hepatitis B virus

The hepatitis B virus is a DNA virus (whereas the hepatitis A virus is an RNA virus) which presents many original features.

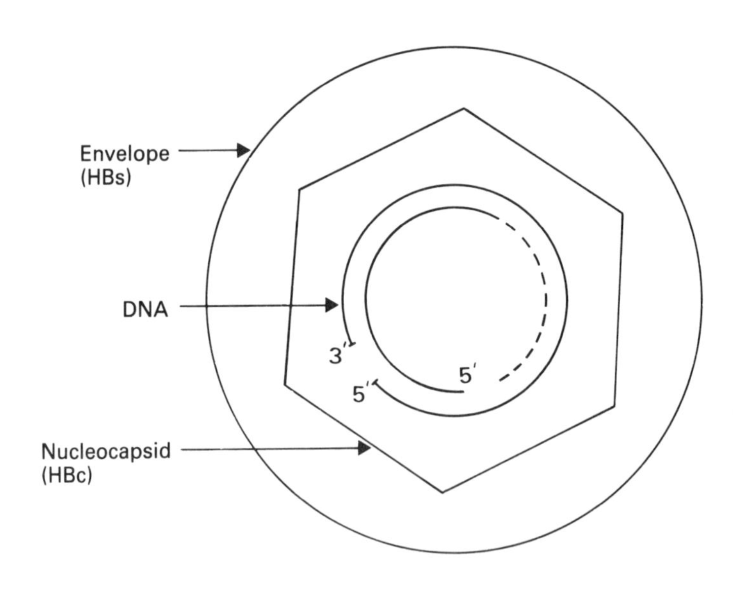

Fig. 107 — Hepatitis B virus.

It is the smallest of the known DNA animal viruses. (As a comparison, the herpes virus, for example, consists of about 100 000 nucleotides.) The hepatitis B virus has a partially double-stranded DNA. In fact it contains:

— a long strand, of constant length (3200 nucleotides). The sequence of this long strand was determined in 1979 (for the sub-type ayw) by the French teams of Galibert and Tiollais working together. This strand is circular but not closed.
— a short strand of variable length

The circular structure of this virus is maintained by the complementarity of the bases at the 5′ end of the two strands over a length of about 200 nucleotides. In addition it comprises:

— a DNA polymerase (which is really a reverse transcriptase,
— a nucleocapsid which contains:
  - The protein carrying the HBc antigen (HB for hepatitis B, c for 'core'). This internal protein, as opposed to the surface protein, closely encloses the genome. It is rich in basic amino acids (positively charged), thus forming ionic bonds with the central nucleic acid (negatively charged).
  - The HBe protein, ('e' simply because the letters a,b,c,d were already being used for this virus — a and d, for example, designate antigenic determinants. As it is not easy to determine the origin of the 'e' suffix in any text, certain people say derisively 'e' for enigmatic! (or 'e' for epidemic.) This HBe protein seems to be a soluble form dissociated from the HBc protein:
— an envelope formed from a double layer of phospholipids containing three proteins of differing length (they have the COOH end in common, but are either longer or shorter at their NH_2 end):
  - a 'large protein' (coded by the pre-S1 and pre-S2 regions and the S gene)
  - a 'medium protein' (coded by the pre-S2 region and the S gene)
  - a short protein called 'major protein' because it is the most abundant in this envelope (coded by the S gene).

These three proteins present an HBs (s for surface) antigenic activity. The structural unit which carries the full HBs antigenic activity is a dimer of two major proteins united by two disulphide bridges.

2. The genes (overlapping!)

The genes of this virus all seem to be localised on the long strand. At first sight it seems strange that a strand as short as 3200 nucleotides could code for all these proteins. In fact, overlapping genes are found in the hepatitis B virus, i.e. they can be read in three different reading frames. Thus there is:

— in the first frame: the C gene coding for the HBc protein; the S gene, preceded by the pre-S_1 and pre-S_2 regions (coding for the three HBs proteins);
— in the second frame: the X region, which codes for a peptide whose sequence, but not its role is known;
— in the third frame: the P region, which overlaps part of the C region, the whole of the S gene and part of the X gene. This region of the DNA probably codes for the viral DNA polymerase.

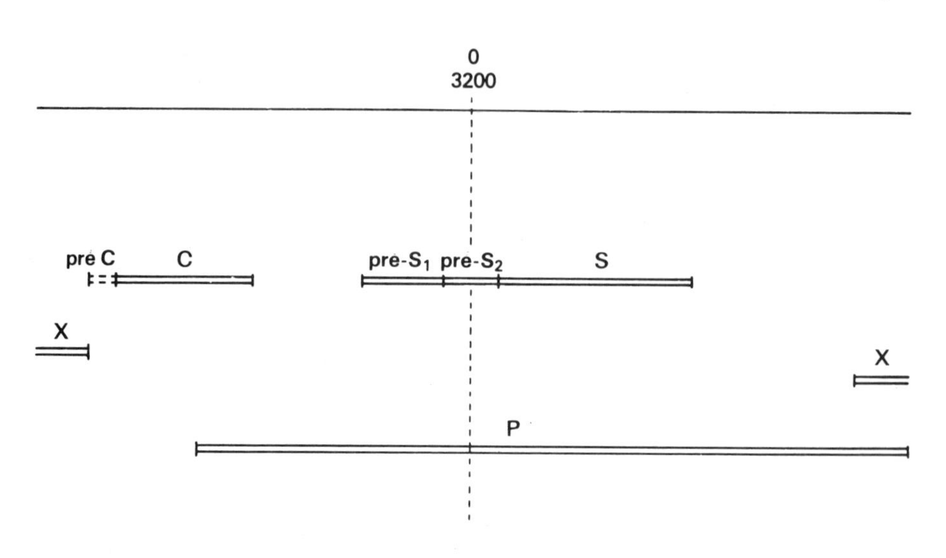

Fig. 108 — The genes (overlapping) of the hepatitis B virus (Representation of the circular genes by a straight line shows that the X gene may be cut in two here).

It is still not known if several mRNAs corresponding to each of these proteins are formed or if a single mRNA is translated in three different reading frames. However, two types of mRNA have been isolated from the liver cells of chimpanzees infected by the virus: one of about 2100 nucleotides, and the other of 3500 nucleotides, so that it is larger than the long DNA strand which only contains 3200 nucleotides! (This arises from the fact that transcription takes place from a circular DNA and can perform more than one turn, which is all the more significant since the total number of nucleotides (3200) is not a multiple of 3!)

3. *Replication of the hepatitis B virus*

The replication of this virus is quite amazing. It has been studied in an allied virus, that of Peking duck hepatitis. Usually the DNA or DNA viruses (as with the DNA of of our own cells) is replicated directly to give new DNA molecules. In the case of the hepatitis B virus, an intermediate form of RNA appears! First of all the virus which penetrates into a hepatocyte is liberated from its envelope and then from its capsid. The viral DNA enters the nucleus and is transcribed into RNA(s) by the RNA polymerase of the hepatocyte. These RNAs can be translated into viral proteins (capsid proteins, envelope proteins, DNA polymerase etc...). They also serve as replication intermediaries.

In fact the RNA obtained (considered as an intermediary in replication) as well as the DNA polymerase will be found again covered by a capsid. The DNA polymerase (which here functions as the reverse transcriptase of retroviruses) recopies the RNA as a single-strand of DNA. The RNA model is hydrolysed, and a second strand of DNA complementary to the first, begins to be synthesised. This complementary copy will not generally have time to be finished. Synthesis is interrupted as soon as the capsid is covered by the envelope. This explains why the second strand may be shorter and why its length may be variable.

It is strange to think, as Tiollais *et al.* have remarked, that the hepatitis B virus reproductive cycle is in a way a mirror image of the reproductive cycle of retroviruses. In fact:

- the hepatitis B virus
  - — is a DNA virus,
  - — passes through an intermediary RNA,
  - — gives a further DNA thanks to a reverse transcriptase;

- a retrovirus
  - — is a RNA virus,
  - — passes through a DNA intermediary thanks to a reverse transcriptase,
  - — gives a further RNA.

IV. COMMENTS ON VIROIDS AND PRIONS

1. Viroids

It has long been thought that nothing simpler than a virus could exist. In fact viroids have been discovered, which are formed from only a simple piece of nucleic acid (RNA) not protected by proteins.

It is their very particular structure (only known since 1978) which allows viroids to resist attacks by nucleases. Here there is an RNA with no free end in the shape of a rod (flattened circle), two thirds of whose nucleotides are paired (by self-complementarity). Such a structure is entirely original and had never been observed in any virus.

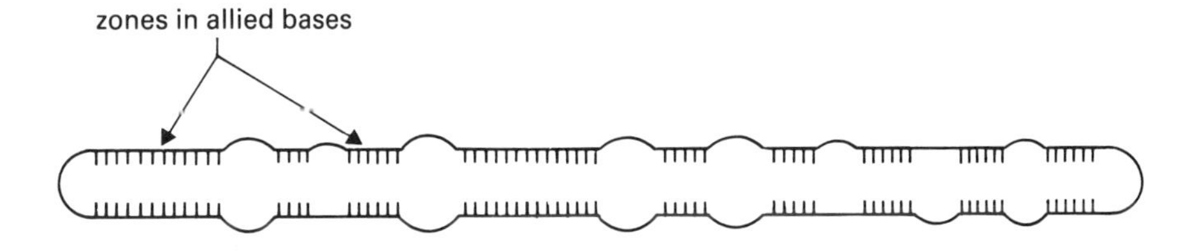

Fig. 109 — Structure of a viroid (circular RNA).

Viroids are responsible for diseases at present known only in plants, but the possibility cannot be excluded that they may also be responsible for certain human diseases.

2. Prions

In 1982 new infectious agents termed prions were detected. Their structure is not yet precisely known. They appear to be deprived of nucleic acid and to contain only a small protein. The reproduction of these prions poses a problem: how could this infectious agent multiply if it is really without nucleic acid? Does it in fact contain a small piece of nucleic acid which has so far gone unnoticed?

Prions are responsible in animals for a disease which attacks the nervous system ('trembling' in sheep and goats).

In man it is suspected that certain diseases of the central nervous system could be connected with prions (transmissible encephalopathies of slow development such as kura or Creutzfeldt–Jakob syndrome).

6

Summary of the various DNA↔RNA conversions described in the preceding chapters

The different conversion operations of DNA↔RNA described in the earlier chapters can now be summarised in the following diagram.

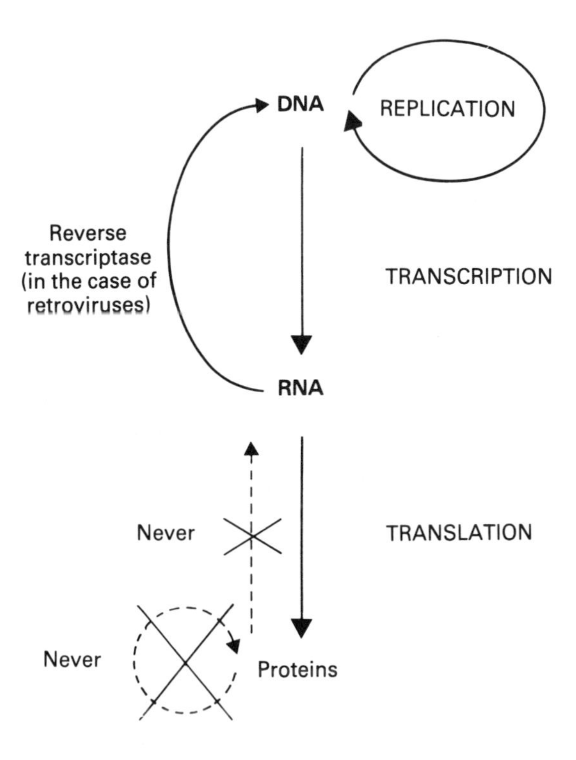

Fig. 110 — Summary of different DNA↔RNA conversions.

7

Cancer

I. DEFINITION

A cancerous cell is a cell whose growth and division are no longer controlled. It therefore divides anarchically, producing millions of cells which will thus constitute a tumour.

In addition cancerous cells lose their specialisation. They are said to 'de-differentiate'.

Finally they can travel distances, producing 'metastases'.

If one studies the transformation of a healthy cell to a malignant cell *in vitro*, the following characteristics are essentially observed (characteristics moreover which are not acquired simultaneously but successively):

- 'immortalisation' (the possibility of a theoretically unlimited number of divisions).
- 'transformation' which is manifested by:

— the loss of 'contact inhibition'. The cells no longer cease their multiplication when they come in contact with each other. They overlap and form small foci consisting of many layers of cells which pile up instead of forming monocellular layers. The form of the cells is also modified, and one can see extensions infiltrating between neighbouring cells.

— the loss of 'the need to anchor'. The cells no longer need to anchor themselves on a solid medium in order to divide.

We shall see that this is largely due to a modification of the properties of cytoskeletal proteins (the internal framework of the cell).

II. ORIGIN OF CANCER IN MAN

Cancer in man can be linked to multiple causes. It can, for example, be related to hormonal imbalance (e.g. certain cancers of the breast), to the environment, such as radiation emitted by the sun (in cancer of the skin), radioactive radiations (in cancers of the blood), tobacco (in lung cancer), and certain chemical products used as food additives (in colonic cancers).

Only certain cancers, which, however, only represent a small percentage of human cancers, are linked to viruses. These are DNA viruses such as, for example, the hepatitis B virus in early liver cancer, the Epstein–Barr virus in Burkitt's lymphoma and rhinoparyngeal tumours, or papillomaviruses in certain cutaneous and genital cancers. Amongst the RNA viruses, a family of human retroviruses has been identified, the HTLV-I and II viruses, connected with the development of leukaemias and of T-cell lymphomas. (These two types of virus should not be confused with the HTLV-III virus — first called LAV in France, and now HIV — the virus responsible for AIDS, which does not belong to the same family as HTLV-I and II.)

It will be seen that the retroviruses which carry 'oncogenes', although capable of initiating numerous cancers in their natural animal host (chicken, mouse, etc.) have not however been accounted responsible for human tumours.

It has been understood for some years that cancer results from stimulated cell division, but the cause of this stimulation (or what prevented the usual control) remained unexplained. In these last years, enormous progress has been made with the discovery of 'proto-oncogenes'.

III. PROTO-ONCOGENES

1. Definition and some examples

In normal conditions we all possess in our DNA (as also do the other mammals and even other eukaryotes) genes which are called proto-oncogenes (from the Greek 'protos', first, or precursor; and 'oncos', tumour). These genes code for proteins which are still not all known, but which probably play a very important physiological role in cell growth and differentiation.

We shall see that certain retroviruses can, when they parasitise the cells of their natural host (mouse, chicken, etc.), accidentally acquire proto-oncogenes. The proto-oncogene will then be able (in conditions still not well understood) to become an oncogene which the virus will transmit finally to an animal host.

Thanks to these retroviruses, it has been possible to study the majority of oncogenes. It is for this reason that now not only the oncogenes, but also the proto-oncogenes, are designated by three-letter abbreviations which recall their initial characterisation. Thus, for example:

— *src*: chicken sarcoma virus (of Rous),
— *ras-ha*: mouse sarcoma virus (of Harvey),
— *ras-ki*: mouse sarcoma virus (of Kirsten),
— *sis*: monkey sarcoma virus (Simian),
— *fes*: feline sarcoma virus,
— *myc*: avian myelocymatosis virus,
— *myb*: avian myeloblastosis virus,
— *yes*: Y73 sarcoma virus of birds,

— *ros*: UR2 virus of chickens,
— *erb*: avian erythroblastosis virus,
— *fps*: Fujinami sarcoma virus of birds,
— *mos*: Maloney sarcoma virus of mice,
— *fos*: FBJ osteosarcoma virus of mice,
— *abl*: Leukaemia virus (Abelson's) of mice, etc.

There are cases where the cellular proto-oncogene has not been identified in a virus. An example of this is a proto-oncogene related to the *ras* family (*N-ras*) found in isolated DNA of different types of human tumour.

The localisation on our chromosomes of many of the proto-oncogenes cited here is now known.

It should not be thought that the proto-oncogenes which we (and other mammals) normally possess in our DNA have been given to us by viruses. It does not prevent the designation of each of these proto-oncogenes referring very clearly to the name of a virus, even if up to now the retrovirus responsible for cancer formation in man has not been found! (which is hardly a help).

Thus it will be seen, for example, that the *ras* proto-oncogene can, in man, be responsible for a cancer, following a modification where no virus whatever has been involved. However, 'ras' is the name taken from the Harvey virus responsible for the transmission of the *ras* oncogenes to the mouse in which it causes a sarcoma (connective tissue cancer).

2. Proto-oncogenes have a cellular, not viral, origin

Some very interesting research has enabled the carrier viral DNA fragment of a viral oncogene (in this case *src*) to be compared with the corresponding carrier DNA fragment of the cellular proto-oncogene (in the chicken). These double-stranded DNAs have been denatured to obtain single-stranded DNAs which can then by hybridised. The viral gene and the cellular gene hybridise, forming a mixed double-stranded molecule which can be observed by electron microscopy. And here is a surprise! Six loops are seen which correspond to six introns contained in the strand of cell DNA, but not contained by the retrovirus gene. These results, fundamentally, indicate that the cell proto-oncogenes originate in the cell and have not been introduced by viruses. Remember moreover how it was previously emphasised that the oncogenes found in retroviruses are not essential to the retroviruses (these are only the *gag*, *pol* and *env* genes). By contrast, the proto-oncogenes would appear to play a fundamental role in man and in other mammals in cell growth and in differentiation.

IV. PROTEINS CODED FOR BY PROTO-ONCOGENES AND ONCOGENES

The action of a proto-oncogene or of an oncogene is obviously due to the protein for which this segment of DNA codes.

When a proto-oncogene has been identified the following questions always arise:

1. For which protein does this proto-oncogene code in normal conditions?

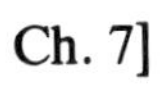

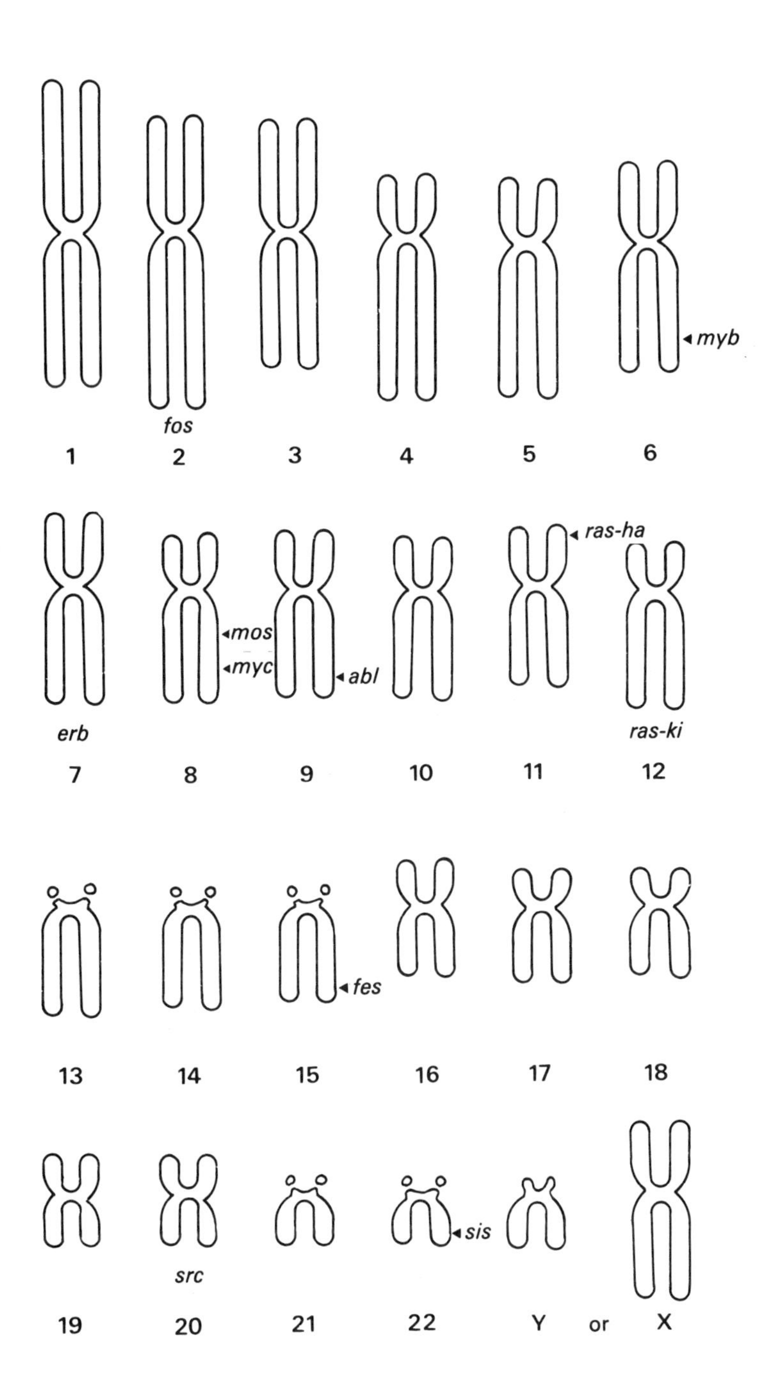

Fig. 111 — Localisation of some proto-oncogenes on human chromosomes.

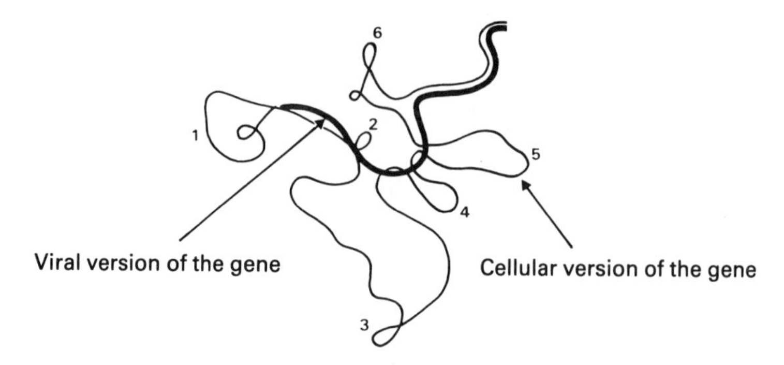

Fig. 112 — Hybridisation of carrier viral DNA carrying the *src* gene and of the chicken DNA carrying the cellular version of the gene. From *Science*, May 1982, p. 34, Fig. 9. Bishop, M., 'The Oncogenes'.

2. What is the physiological role of the protein coded for by the proto-oncogene?
3. What sort of modification (qualitative and/or quantitative) can transform a normal protein into an oncogenic protein (or translation of a qualitative and/or quantitative modification of the proto-oncogene)?

In order to try to simplify the explanation, we shall regroup some of the known proto-oncogenes/oncogenes into families.

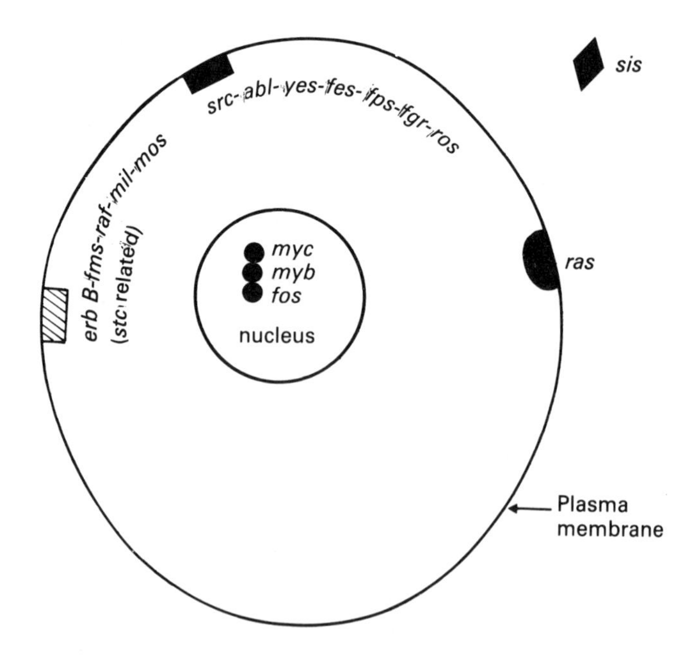

Fig. 113 — Localisation of the proteins coded for by the different families of proto-oncogenes/oncogenes.

A. The *ras* family

1. The ras *proto-oncogene*

The protein coded for by the *ras* proto-oncogene is a membrane protein (situated on the inner surface of the plasma membrane) of molecular weight 21 000, from which comes its name 'p21'. Its exact role is still not known. It can merely be stated that it forms part of the G protein family: it has the property of binding the GTP and then of hydrolysing this GTP to GDP. It is possible that a short cellular signal is transmitted inside the cell during the short space of time that the p21 protein is linked to GTP.

2. The ras *oncogene*

A mutation of a base plays a role in the transformation of the *ras* proto-oncogene into an oncogene. How has it been shown that the segment of DNA where this mutation occurs was endowed with oncogenic properties?

(a) Detection by the transfection test

A technique known as a 'transfection' test has been developed to determine whether a DNA segment contains an oncogene. It consists of causing DNA segments originating from tumour cells to penetrate certain cells (offspring of mouse fibroblasts termed 3T3 cells) and of determining which DNA segment will provoke the transformation of the 3T3 cells.

The 3T3 cells used in this test are very special cells. They are in fact more sensitive than the normal transformation cells (since they have already gone through immortalisation, the first phase towards cancerisation).

Thanks to this test, it has been possible to identify the fragment of gene (coming from the DNA of tumour cells) responsible for the oncogenic properties of *ras*.

(b) G→T mutation in the ras *oncogenes, and Gly→Val in the* ras *oncogene protein*

The nucleotide sequence of the oncogenic DNA segment has been compared with that of the corresponding proto-oncogene. One qualitative anomaly was revealed. This is a simple mutation where codon 12 has undergone the replacement of a base. Thus:

ras proto-oncogene ... 5′GCC GGC GGT ... 3′
(codes for Gly)

ras oncogene ... CGG GTC GGT...
(codes for Val)

For clarity, only the non-transcribed DNA strand (the strand of DNA complementary to the transcribed strand, and thus comparable in its base sequence to mRNA) has been represented here.

The replacement of G by T brings about the replacement of Gly by Val in the part of the protein coded by the *ras* oncogene fragment.

(c) Determination of mutations

- Cellular progeny

This mutation transforming the *ras* proto-oncogene into the *ras* oncogene was first found in 1982 in the DNA of laboratory-cultured cells from a human bladder cancer (cellular EJ progeny of the team of Weinberg, and T24 of the team of Barbacid).

- Human tumours

These first results were very enthusiastically welcomed by the scientific community. However, given that they had been obtained on cellular progeny, the possibility remained that the mutations could have been produced *in vitro*. Since then, these mutations have been found directly in human tumours. It is in this way that oncogenes coding for Ser (instead of Gly) have been identified in colonic, pulmonary and pancreatic tumours. Nevertheless, they were only found in about 20% of the cancers studied.

Moreover, it was equally interesting to confirm that the proto-oncogene (coding for Gly) was found not only in healthy subjects but also in healthy cells (sampled in tissues remote from the tumour) in the same cancerous subjects.

(d) Activation of the ras *proto-oncogene* in vivo

Even if *ras* proto-oncogenes were capable of transforming certain cells (3T3 cells) *in vitro*, and had been found in human tumours, it had not been proved that a proto-oncogene could be transformed into an oncogene *in vivo* under the influence of a carcinogenic product. This is now an established fact. In May 1985, Barbacid's team showed that the *ras* proto-oncogene could be activated into an oncogene *in vivo*, by the chemical carcinogen NMU (*N*-nitroso-*N*-methylate). This product induces mammarian cancers in the rat (after a simple exposure during the period of sexual development). NMU is responsible for G→A transitions. Thus, it transforms the *ras* proto-oncogene at the critical codon 12 'GGA' (which codes for Gly) into 'GAA' (which codes for Glu). This oncogene is easily detected in mammarian tumours induced by NMU.

(e) Distinctions between the origin of human and animal tumours

It is very important to understand that in man, mutations responsible for the replacement of Gly by another amino acid can be provoked by exposure to radiation or to a chemical carcinogen, but not, it would seen at present, by the presence of a virus.

On the other hand, in mice or rats, it is not only radiation or carcinogenic products, but also viruses which can be responsible for the appearance of oncogenes in the host cell DNA, thus provoking a tumour. It has been shown that the oncogenes transmitted by viruses code for proteins where Gly will be replaced by another amino acid. Thus, the Harvey virus will supply the *ras-ha* oncogene (which will code for Arg), the Kirsten virus the *ras-ki* (which will code for Ser) and the virus of the BALB murine sarcoma the base oncogene (which will code for Lys).

As for normal rat cells, they possess the same *ras* proto-oncogene as man, coding, therefore for Gly.

(f) The ras protein–GTP linkage

Thus it seems that the transforming action of the protein coded by the *ras* oncogene stems from the fact that Gly (in the 12th position in the normal p21 *ras* protein) is replaced by another amino acid.

Gly is a very special amino acid, by virtue of the absence of a residue carried by the α carbon. Together with Pro it is the greatest known breaker of the α helix. It is probable that the protein having an amino acid other than Gly at this precise area of the molecule has a different spatial form and links in a different way to GTP. In fact, it has been determined that the oncogenic p21 *ras* protein–GTP link is more stable than the normal p21 protein–GTP link.

This greater stability has also been explained, through the discovery that the oncogenic p21 protein has lesser GTPasic properties than those of the normal p21 protein.

The fact that the oncogenic p21 protein–GTP linkage is more stable has enabled us to advance various hypotheses for this protein's mode of action. For example, a signal for cell division will be transmitted by p21–GTP intermittently in the case of normal p21 (since GTP is rapidly hydrolysed into GDP) and permanently in the case of oncogenic p21. However, these are still only hypotheses.

(g) How is the cellular ras *proto-oncogene transformed into a* ras *oncogene in the virus*

If a retrovirus is not responsible in man for the transformation of the *ras* proto-oncogene into a *ras* oncogene, the question obviously arises (but only to satisfy our intellectual curiosity) as to why, or rather how, the cellular proto-oncogene accidentally acquired by the retrovirus becomes an oncogene for the animal. In fact, as will be seen, retroviruses pick out cellular 'proto-oncogenes' in the habitual host animals (mouse or rat) and then transmit 'oncogenes' to healthy animals which they subsequently infect.

Firstly, remember that retroviruses are RNA viruses which multiply through the intermediation of DNA. This DNA, formed by the action of reverse transcriptase, can then be integrated into the DNA of the host cell. It is the RNA polymerase of the host cell which will transcribe the viral DNA. The mRNAs will be transported in the cytoplasm. One part of these mRNAs will be translated into viral proteins, another part will constitute the viral genome destined to be incorporated with viral proteins to form new virions. One can distinguish between retroviruses which do not possess an '*onc*' gene, which are the more numerous (e.g. the AIDS virus) and the retroviruses called 'oncogenes' which do possess them and are much more rare. These *onc* genes are not essential to these viruses and can be considered to some extent as parasites(!). We have seen that genes indispensable to a retrovirus are the *gag*, *pol* and *env* genes.

How then does one explain transformation, in the virus, of the accidentally acquired cellular *ras* proto-oncogene into a *ras* oncogene? Different hypotheses have been put forward. It could be, for example, a mutation (at the triplet normally coding for Gly). It is not yet exactly known if it is a 'copying accident' arising at the moment of transcription of 'integrated DNA→viral RNA' or if this mutation is produced later in the virus. It is no longer possible for any other mechanism to be involved.

In fact we shall see that possibilities exist other than *qualitative* modifications to render a proto-oncogene 'oncogenic'. These are *quantitative* modifications or 'over-expression'.

(h) Conclusion: activation of the ras *proto-oncogenes occurs by different mechanisms in man and animal*

What, finally, is truly remarkable, is that two completely different mechanisms may activate the same proto-oncogene (initially found in both a human cell and a rat cell) mutation (provoked by radiation or a carcinogenic chemical substance) in man; acquisition from a retrovirus in the rat.

B. The *src* family

The *arc*, *abl*, *yes*, *fes*, *fps*, *fgr* and *ros* proto-oncogenes can be grouped into the same family.

1. Proteins coded for by src genes have tyrosine kinase properties

The proteins coded by *src* and *abl* proto-oncogenes are the best known. They have molecular weights of 60 000 and 120 000 respectively from which they get their names 'p60' and 'p120'. The proteins coded by the proto-oncogenes of this family are kinases which have the special characteristic of phosphorylating Tyr. The kinases known previously phosphorylate Ser (and perhaps Thr) but not Tyr. Which then are the proteins whose Tyr is phosphorylated by tyrosine kinase? It has been possible to identify proteins situated on the inner surface of the plasma membrane such as the p36 protein, glycolytic enzymes and also cytoskeletal proteins such as vinculin. Vinculin is found in attachment plates which affix the culture cells to surfaces. Although vinculin contains many Tyr residues, a sole Tyr is phosphorylated in the normal state. Vinculin then is normally weakly phosphorylated by the proteins coded by the proto-oncogenes of the *src* family (in particular by *src* p60 and *abl* p120).

In unusual physiological conditions, i.e. in transformed cells, (cells that are no longer normal), the phosphotyrosine concentration is seen to be increased. The question then arises as to the action of these type p60 and p120 proteins. Do they provoke phosphorylation of other tyrosines on vinculin, or on some other proteins of the cytoskeleton (normally not phosphorylated), thus bringing about disorganisation of the cytoskeleton?

2. The coded proteins can be hybrids

It is strange to note that proteins coded for by virus oncogenes of this family (with the exception of *src* however) are hybrid proteins. In effect, they contain a part (NH_2 side) coded for by the viral gene *gag* and another part (COOH side) coded for by the oncogene responsible for the 'tyrosine kinase' properties (*abl*, *fes*, *fps*, *fgr* or *ros*). The protein coded by viral oncogene *abl* is an example.

The protein coded for by the *src* oncogene on the other hand, only possesses the region involved in the tyrosine kinase properties. It is interesting to note that there is a great resemblance in the amino acid sequence of the tyrosine kinase regions of all these proteins.

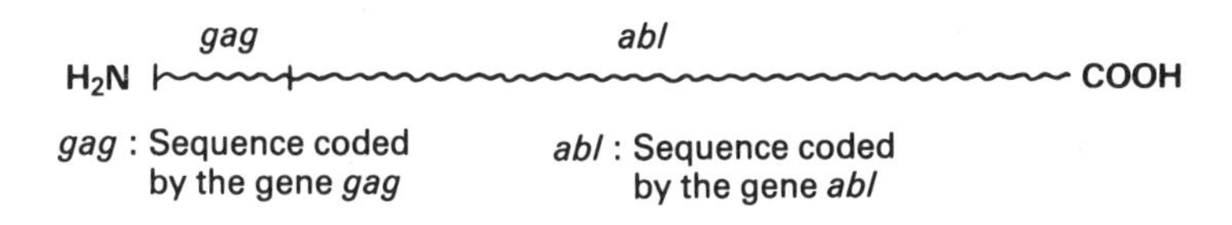

Fig. 114 — The oncogenic protein coded by the *abl* virus is a hybrid protein.

C. Family related to the *src* family

1. Relationship between *src* proteins and *erb-B* proteins, etc.

Another family of proto-oncogenes (*erb-B*, *fms*, *raf*, *mil* and *mos*) is related to the *src* family. In fact the proteins coded for by the proto-oncogenes of this family possess, in common with the proteins coded for by the proto-oncogenes of the *src* family, the sequence situated at the COOH extremity. Nevertheless these proteins do not have any tyrosine–kinase activity.

2. Comparison of the protein coded for by *erb-B* and the EGF receptor

Recent research has shown that the protein coded for by the *erb-B* oncogene

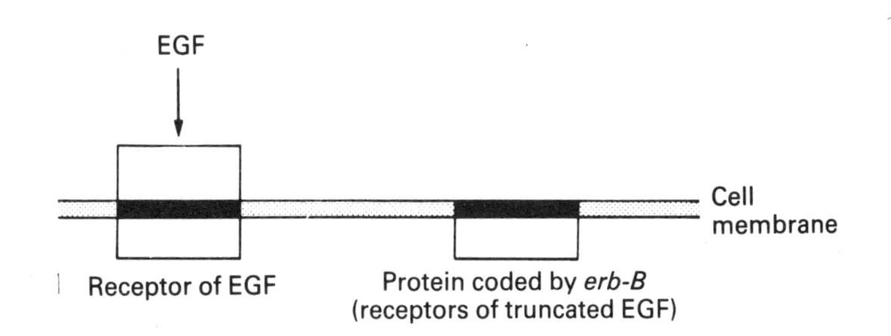

Fig. 115 — The protein coded for by *erb-B* appears to be a truncated EGF receptor.

manifests a very strong relationship with another protein ('the epidermal growth factor' — EGF). The receptor for this protein contains (1) an extracellular region which fixes the EGF, (2) a transmembranous region, and (3) an intracellular region. The protein coded for by *erb-B* appears as a truncated EGF receptor. It contains in fact areas 2 and 3 (transmembranous and intracellular).

D. The '*myc*' family

The *myc*, *myb* and *fos* proto-oncogenes code for nuclear proteins which interact directly with DNA and are perhaps also inclined to increase replication.

The sequence of the *myc* proto-oncogene has recently been compared with that of the *myc* oncogene and some differences observed.

E. The '*sis*' family

1. *The protein coded for by the* sis *proto-oncogene is secreted*

The *sis* proto-oncogene forms a separate family. The protein coded by the *cis* proto-oncogene is neither membranous (like the proteins coded for by genes of the *ras* and *src* families), nor nuclear (like those coded for by genes of the *myc* family), but it is secreted.

2. *The protein coded for by the viral* sis *oncogene is a hybrid protein*

First remember that there is a virus possessing the *sis* oncogene responsible for the monkey sarcoma. The *sis* oncogene differs qualitatively from the cellular proto-oncogene. Point mutations are found for example. In addition, it seems that the protein coded by the virus is a hybrid protein. This is due to the fact that one part of the protein coded for by the viral *env* gene was synthesised with the oncogenic protein into a single product. This is no exception. We have seen, with the *src* family, examples where a part of the protein was coded for by the viral *gag* gene.

3. *Analogies between the sequence of the* sis *protein and PDGF*

A disturbing discovery concerning the protein coded by the *sis* oncogene was made in 1983. The amino acid sequence of this protein is very close to that of a growth factor found normally in human blood platelets — PDGF ('platelet-derived growth factor').

The PDGF receptor is a protein (situated on the plasma membrane) whose tyrosine kinase properties are stimulated during fixation of PDGF. The protein coded for by the *sis* oncogene, which has a similar structure to that of PDGF, is also capable of fixing itself on the PDGF receptor and probably also capable of activating the tyrosine kinase properties of this receptor.

4. *Tyrosine plays a key role in the control of growth*

Thus it appears that Tyr plays a key role in the control of growth. It is strange to state that a growth factor (like PDGF), which stimulates a controlled cellular growth, and a tumour virus (like the monkey sarcoma virus), which provokes uncontrolled growth, both act by activation of a kinase phosphorylating Tyr. It could be supposed that a signal responsible for cellular division — initiated by phosphorylation of Tyr — would be transitory in one case (normal growth) while in the other case (cancer) it would be permanent. This reminds us of what has already been revealed about the role of *ras* proteins which bind GTP and would themselves provoke an intermittent or permanent signal.

V. SOME EXAMPLES OF TRANSFORMATIONS OF PROTO-ONCOGENES INTO ONCOGENES INVOLVED IN HUMAN TUMOURS

Let us take some examples in man of possibilities for transformation of a proto-oncogene into an oncogene (possibilities which are, however, far from all being known).

Two types of modification, qualitative or quantitative, are possible at the level of proteins coded for by oncogenes. These two types of modification can occur simultaneously.

A. Qualitative modifications: point mutations

The example of the *ras* family has shown that a qualitative modification (point mutation) could be responsible for the transformation of a proto-oncogene into an oncogene and that the protein coded by the *ras* oncogene had modified properties (bonding with GTP). It cannot be excluded that in this case a quantitative modification of coded proteins may also be found.

B. Quantitative modifications: over-expression

In general there are only two examples of proto-oncogenes in the cell producing a fixed quantity of mRNA and proteins. In some cases an increase in the number of transcribed mRNA molecules and translated proteins is seen; this is 'over-expression'. This then would be a quantitative (rather than qualitative) modification which would here be responsible for malignant transformation. However, why might a protein necessary for the normal functioning of the cell, have carcinogenic effects when it is synthesised in excess? Over-expression can be obtained in various ways.

1. Genetic amplification

Over-expression can be due to the existence, in the DNA of a cell, of multiple copies of a proto-oncogene.

Thus, in a cellular progeny (HL 60) derived from an acute human myeloid leukaemia, multiple copies of the *myc* proto-oncogene have been found.

2. Activation of gene transcription

The increase in the activity of transforming proteins can be due to transcription activation. This activation can be explained essentially in two different ways.

(a) Insertion of a viral promoter

It is possible that certain human cancers may be linked to the fact that a virus, although deprived of oncogene, is incorporated into the DNA of a cell, by chance in the proximity of a proto-oncogene. The proto-oncogene then comes under the control of the viral promoter and is actively transcribed. The virus, deprived of oncogene, in this case acts solely through the mediation of its promoter. The cancer would be explained here not by the presence of an abnormal protein but by the over-expression of a normal protein. The virus would not therefore be directly responsible for this over-expression. Such a mechanism is known for the ALV (avian leukaemia virus). This virus has no oncogene, but it is nevertheless capable of provoking a B-cell lymphoma in the animal. In fact, activation of the cellular *myc* proto-oncogene is produced by insertion of the viral promoter in the vicinity. The *myc* gene which is then under the control of the viral promoter is actively transcribed.

(b) Chromosomic translocations

- Displacement of a proto-oncogene towards a zone of active transcription

Activation of transcription can be due to a displacement of a proto-oncogene situated in an inactive zone or a zone of little activity towards a zone where the genes are actively transcribed. This proto-oncogene displacement can occur from one chromosome to another! An example of an active transcription zone is that of regions coding for immunoglobulins in B lymphocytes. These zones are found on chromosomes 14 (for the synthesis of heavy chains), 2 (for light kappa chains) and 22 (for light lambda chains).

- Role of 'enhancers'

Why is the expression of genes so active in regions coding for immunoglobulins? In certain cases it is perhaps because, in cells capable of synthesising immunoglobulins, there are sequences called 'enhancers' (which stimulate the expression of genes as we have already seen). These sequences only function in the specialised cells in the synthesis of immunoglobulins, but one the other hand, they are capable of activating any gene situated in their vicinity in these cells. This is how the proto-oncogenes, placed by accident in the proximity of these 'enhancers', will be activated. In the majority of translocations, it is the proto-oncogene which is placed near a stimulator sequence (these are the examples which are used below). More rarely, the proto-oncogene remains in place and it is the activator sequence which moves. It remains to be seen by what mechanism 'enhancers' increase transcription.

- Examples of chromosomic translocations

— Burkitt's lymphoma

In Burkitt's lymphoma (cancer of lymphoid tissue, frequent in African children and adolescents, but much more rare in Europe), the region containing the *myc* proto-oncogene (on chromosome 8) is exchanged, or 'translocated' with a region situated on chromosome 14 (75% of cases of Burkitt's lymphoma). The *myc* proto-oncogene is now found in the proximity of the gene coding for heavy immunoglobulin chains. It can be observed that after translocation, the chromosomes have a different length from that of their undamaged homologues. (Thus, chromosome 8 is shortened, whereas chromosome 14 is lengthened).

Translocation can also be effected with chromosomes 2 (9%) or 22 (16% of cases of Burkitt's lymphoma).

As we said earlier, the role of the protein coded by *myc* is unknown. It is only known that this protein will act at the level of the cell nucleus.

The association between the Epstein–Barr virus (which is a DNA virus and not a retrovirus) and Burkitt's lymphoma is well known. However, what role does this virus play? It can only be that of an initiating factor, particularly implicated in 'immortalisation' of a large number of B lymphocytes. In fact, this virus can immortalise the multiplication of these cells *in vitro* (moreover these properties are exploited to obtain antibody-producing cellular progeny). It is presently thought that the Epstein–Barr virus can, in stimulating lymphocyte proliferation, increase the frequency of chromosomic translocations.

— Chronic myeloid leukaemia (CML)

In CML, the region of chromosome 9 containing the proto-oncogene '*abl*' is translocated onto chromosome 22 (while the proto-oncogene '*sis*' initially situated on chromosome 22 is translocated onto chromosome 9). Chromosome 22 is finally found to be shortened. This is the famous 'Philadelphia chromosome', the usual marker for CML, recognised since 1960. Many aspects remain to

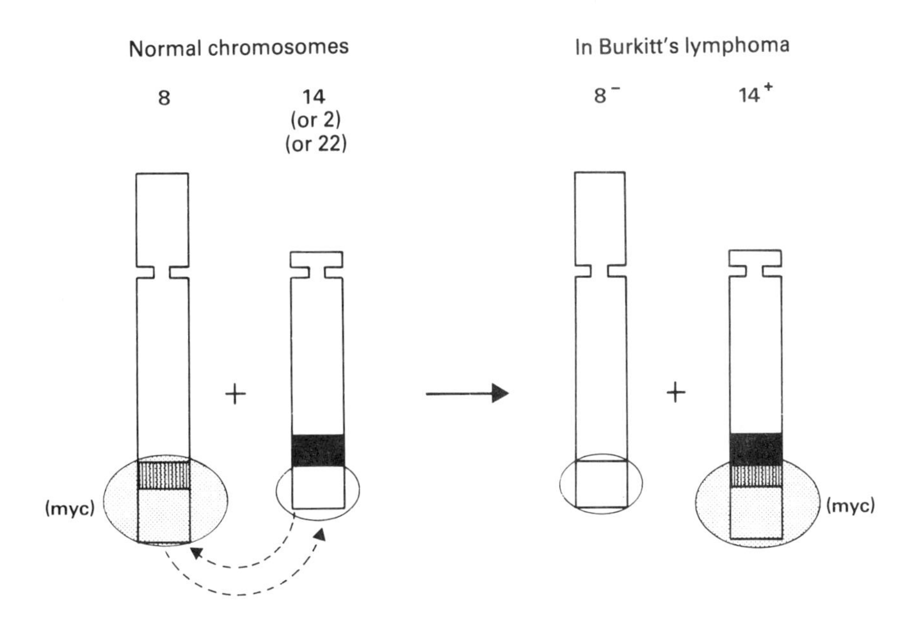

Fig. 116 — An example of chromosomic translocation (observed in Burkitt's lymphoma).

be determined. For example, is there really an over-production of proteins coded by the proto-oncogene '*abl*' because of its translocation onto chromosome 22?

It has been possible to observe other examples of reciprocal translocations. For example, translocation involving on the one hand chromosomes 11, 18 (or others) and chromosome 14 (always 14) on the other. These have been described in B-lymphocyte lymphomas in adults, B-lymphocyte chronic leukaemias, myelomas and also in T-lymphocyte cancers.

It must finally be concluded that although examples of over-expression have been found (due to a genetic amplification or to an increase in transcription) in cases of cancer, it has not been proved that these over-expressions are actually responsible for these cancers! Conversely, over-expressions have been observed in normal cells. It therefore appears that over-expression of a proto-oncogene is not sufficient to provoke cancer.

Summary

Much progress has been made in the knowledge of certain molecular disturbances originating from the transformation of a normal cell into a cancerous cell. Anomalies in proteins coded by oncogenes can be qualitative and/or quantitative. Great progress will be made when the sequence and the role of all the proteins coded by proto-oncogenes and oncogenes is known.

C. Possible involvement of several oncogenes

Recent research has shown that a single anomaly would not be sufficient to obtain a cellular transformation. It seems that at least two events must occur. Thus, in normal cells (this time embryonic fibroblasts of the rat rather than 3T3 cells which have already undergone one stage towards cancerisation) incubated with *myc* or *ras* oncogenes separately there is no transformation, but if incubation is carried out with these two oncogenes simultaneously, a transformation is obtained. *Myc* and *ras* must act according to different and complementary mechanisms. We have seen that *myc* coded for a nuclear protein whereas *ras* coded for a protein localised in the plasma membrane. For many years there had been the conviction that malignant transformation was not a single-factor disease and took place in several stages.

The extraordinary enthusiasm experienced in the light of recent discoveries must, however, be tempered with caution. If it is now possible to advance certain hypotheses based on serious experiments, care must be taken not to generalise the observations obtained with one type of oncogene.

VI. HOW TO DETERMINE IF A PRODUCT IS CARCINOGENIC

Any new synthetic product, whether a drug, a chemical product used in the food industry, or a cosmetic, must be tested with a view to investigating possible cancer-inducing properties.

A. Carcinogenic power in animals

These techniques consist of testing to see if a product is likely to provoke a cancer in the animal. At present they are being replaced by newer, much more simple tests.

B. Mutagenic power on a bacterium

Instead of studying the carcinogenic power of a product on an animal, its mutagenic power is tested on a bacterium. This test is based on the fact that carcinogenic substances are also mutagenic in 90% of cases. The advantage of these tests is that they are quick (taking less than a day, whereas tests on animals can require two to three years), and are much less onerous.

Ames' mutatest is an example. A species of bacterium (*Salmonella typhimurium*) which normally synthesises histidine is used for this test (remember that histidine is an essential amino acid in man, but not in bacteria).

In this type of bacterium, mutants are chosen which are themselves incapable of synthesising histidine. If these mutants are cultured on a medium without histidine, they obviously will not multiply. However, for certain bacteria a spontaneous mutation will occur provoking a return to the initial state (the state where the bacterium is capable of synthesising histidine). This is called a 'reverse mutation'. Thus, if mutants are cultured in a medium without histidine, these mutants will not multiply, except for the reverse mutants: typically about thirty spontaneous reverse mutation colonies will be found.

If a piece of paper steeped in a mutagenic substance is now placed in the centre of the bacterial culture, a much greater number of spontaneous reverse mutations (e.g. 100 to 1000) will be produced.

Before a new product may be administered to a human, other tests should also be

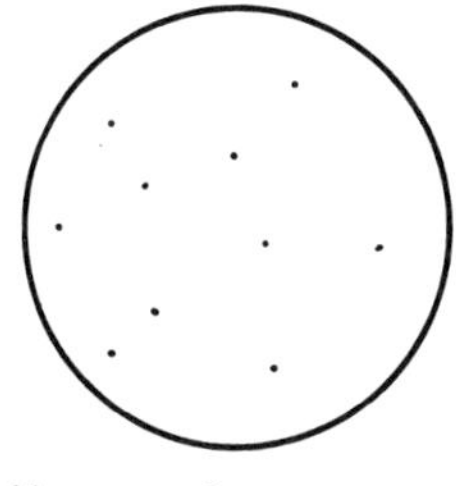

Absence of mutagenic substance (some colonies of 'reverse mutants')

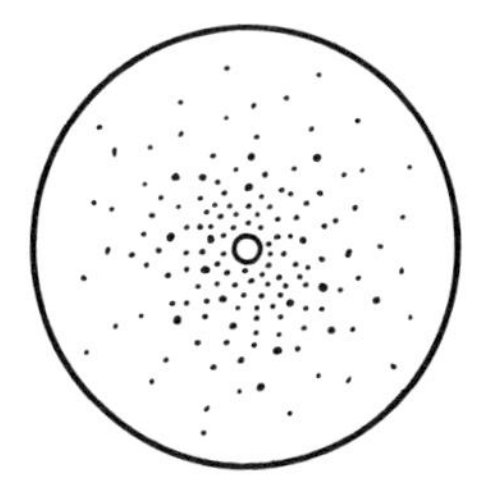

Presence of a mutagenic substance (increase in number of colonies of 'reverse mutants')

Fig. 117 — Testing mutagenic power of a substance by the mutatest.

carried out. In 1984 the European Community recommended a minimum of four tests: a test of genetic mutation on bacterium (Ames-type mutatest) as well as on eukaryote cells, and a test of chromosomal mutation (chromosome modification) *in vitro*, and *in vivo*.

8

Drugs interfering with replication and/or synthesis

In this chapter only schematic ideas as to the mechanism of action of these drugs will be given.

I. ANTINEOPLASTIC AGENTS

Examples of alkylant drugs (Endoxan)

As already mentioned, cancer is due to an exaggerated, anarchic, uncontrolled multiplication of cells. An example of an antineoplastic agent is 'Endoxan', a drug which blocks cell division by blocking replication. One way of preventing replication is to prevent the two strands of DNA from diverging. This is how drugs of the alkylating group, to which Endoxan belongs, will act. An alkyl is a radical obtained by the elimination of OH from alcohol. (For example, $CH_3{-}CH_2{-}$ is a radical alkyl coming from ethyl alcohol). A substance capable of introducing one or more alkyl chains into a molecule is called an 'alkylant'. Antineoplastic agents belonging to the group of alkylants are in fact bifunctional and the general formula of these products is:

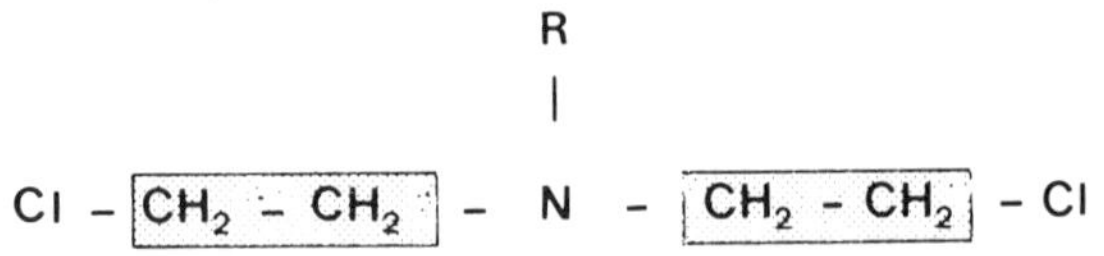

$$
Cl - \boxed{CH_2 - CH_2} - \overset{\overset{\displaystyle R}{|}}{N} - \boxed{CH_2 - CH_2} - Cl
$$

Fig. 118 — General formula for antineoplastic agents belonging to the group of bifunctional alkylants.

The formula for Endoxan for example is shown in Fig. 119.

Alkylants can act at different points in DNA, but the preferred position for alkylation is the nitrogen at position 7 of the pentagonal nucleus of guanine. The alkylant agent (bivalent) creates a bridge between two neighbouring guanines on the same DNA, or on two guanines situated almost face to face on two opposed strands

O NH
P=O
|
Cl - CH_2-CH_2 — N — CH_2-CH_2 - Cl

Fig. 119 — Formula for Endoxan (cyclophosphamide).

of DNA. In this latter case Endoxan forms a bridge between the two DNA chains. In normal DNA, opposite bases are linked by weak hydrogen bonds. At the time of replication these weak bonds allow the two DNA strands to separate easily. This is no longer the case when an alkylant drug creates a bridge (forming covalent bonds) between two strands of DNA. Replication is blocked.

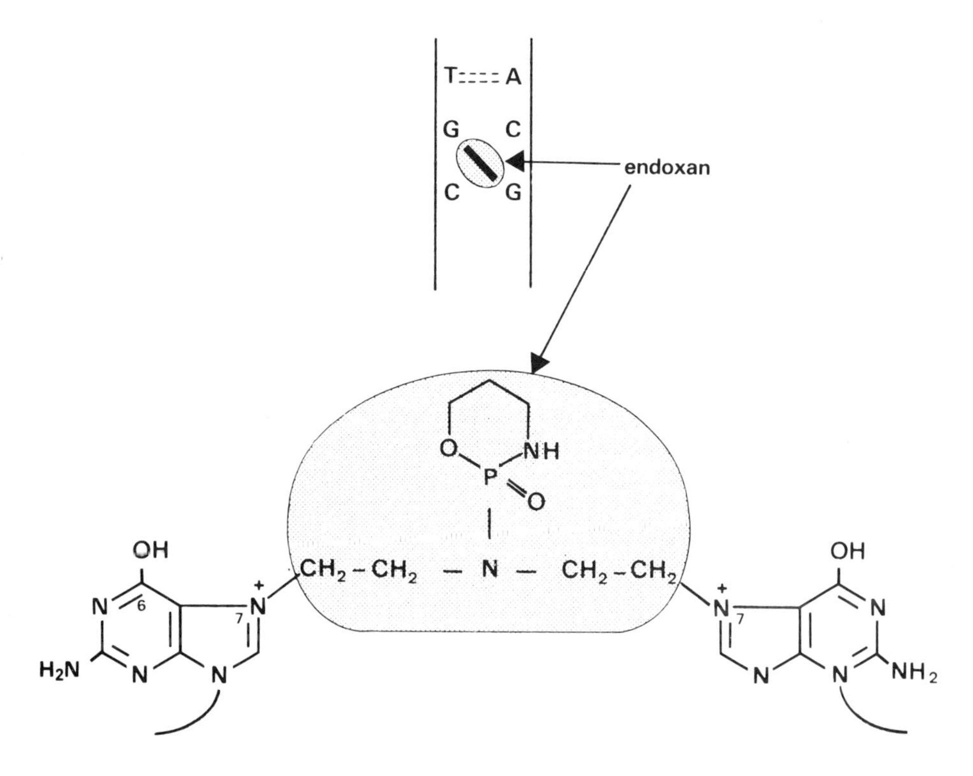

Fig. 120 — Endoxan creates a bridge between two guanines. (Because of alkylation, the nitrogen at position 7 of the guanine becomes a more acidic quaternary ammonium function which facilitates, at position 6, the enolic tautomeric form).

Unfortunately, these agents are not selective for malignant cells and act on all rapidly dividing cells. They can thus block the replication of cells such as those of the bone marrow which synthesise white and red blood cells, resulting in a serious reduction in these blood elements (pancytopenia).

Other ways of preventing DNA replication (e.g. metabolites, etc.) are detailed below.

II. ANTIBIOTICS

Definition

'Bios' signifies life, whence comes the name 'antibiotics' given to drugs designed to fight against life (bacterial, not human).

We are not talking here of those antibiotics utilised in a research laboratory to block a particular stage of protein synthesis, but only (and very briefly) of antibiotics used therapeutically.

The mode of action of these drugs can be very different from one class of antibiotics to another. (Penicillins, for example, inhibit a transpeptidase enzyme which is necessary for the synthesis of the bacterial surface.

- The example of streptomycin

Streptomycin and other antibiotics target the bacterial ribosome. They interfere with translation and prevent the bacterium from synthesising its proteins. Thus, streptomycin causes errors in reading the code, which provokes the incorporation of different amino acids, resulting in an aberrant protein unusable by the bacterium which will finally die.

As opposed to antineoplastic drugs, which cannot distinguish malignant cells from rapidly proliferating normal cells, antibiotics act relatively selectively. In fact, when their mode of action takes place at the level of the bacterial ribosomes, they do not in principle effect the ribosomes of human cells whose structure, as already noted, is a little different from that of prokaryote ribosomes.

III. ANTIVIRAL AGENTS

1. Some examples of antiviral agents inhibiting viral RNA or DNA polymerases

In the struggle against viruses, there is little to go on. There is a tendency to study products inhibiting enzymes which are indispensable to the virus. For this reason agents such as Virustat have been proposed, which will inhibit the RNA polymerase of the influenza virus, Zovirax which will inhibit the DNA polymerase of the herpes virus and HPA 23 (ammonium antimonytungstate), a product still being tested which will inhibit the reverse transcriptase of the AIDS virus, etc.

Although there are relatively few antiviral agents, vaccination against viruses with older vaccines (such as those used to prevent smallpox, poliomyelitis, etc.) has been shown to be very useful. Vaccination represents a future means to combat viral diseases, preventatively and effectively and attempts are being made to prepare new vaccines. Thus a vaccine against the hepatitis B virus has been obtained in recent years and is presently being used in France to vaccinate people at risk of the disease. Active research is also going on to find a vaccine against the AIDS virus.

2. Interferons

Many hopes have centred on interferons in recent years. Interferons represent a family of proteins (MW 15 000 to 35 000) synthesised by cells such as leukocytes and fibroblasts, in response to a viral infection. This synthesis is explained by a derepression provoked by the presence of the virus. Interferons are thus secreted and interact with neighbouring cells. Interferons will protect the neighbouring cells, not by preventing the viruses from penetrating these cells but by preventing their replication; from whence came the name 'interferons' (interfere with virus replication).

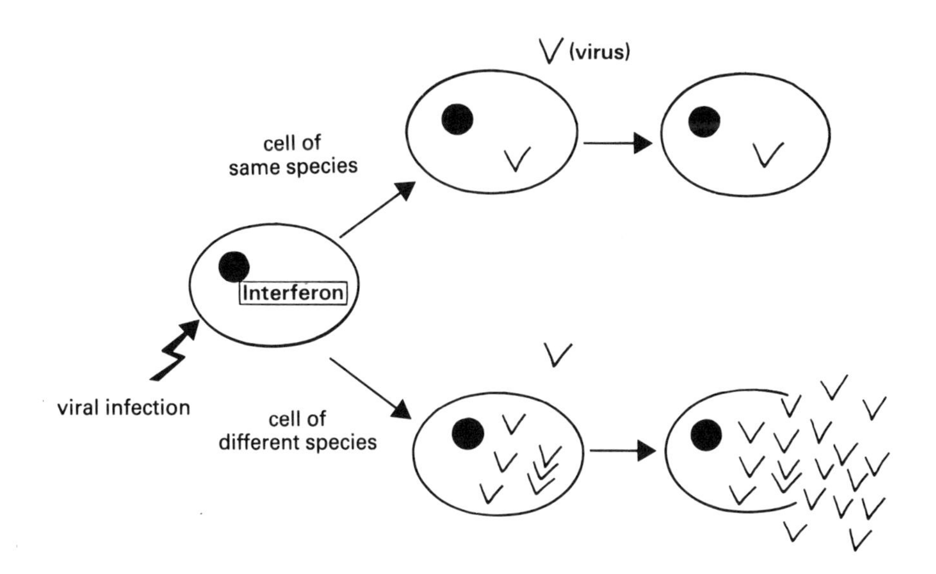

Fig. 121 — Diagram depicting the mode of action and specificity of interferons.

(a) Mode of action

The mode of action of interferons consists essentially of two mechanisms which aim at blocking the synthesis of viral proteins (certain proteins being indispensable to the replication of the virus).

- Inactivation of eIF-2

Interferon is responsible for blockage of protein synthesis during translation (at the initiation stage). In fact interferon provokes phosphorylation (by a protein kinase) of the initiation factor (eIF-2). Once phosphorylated, eIF-2 is inactivated, which will block translation of viral mRNAs.

- Hydrolysis of viral mRNAs

Interferon also provokes the destruction of the mRNAs of the virus. As the transcription of mRNAs proceeds, they are hydrolysed according to the following mechanism. Interferon induces synthesis of an enzyme, the (2′–5′) A_n synthetase. After activation by double-stranded RNAs — these being generated during the multiplication of numerous viruses — this synthetase provokes (from ATP) the formation of an oligonucleotide, a small grouping of pppA(2′–5′)A_n adenine nucleotides (where n=3,4,5 and even more). It is noteworthy that these nucleotides are connected to each other by 2′–5′ bonds and not by 3′–5′ bonds as is usually the case in nucleic acids. (Thus they are resistant to the majority of cellular nucleases!) These oligonucleotides (2′–5′) will finally activate a ribonuclease which will then degrade the viral mRNA.

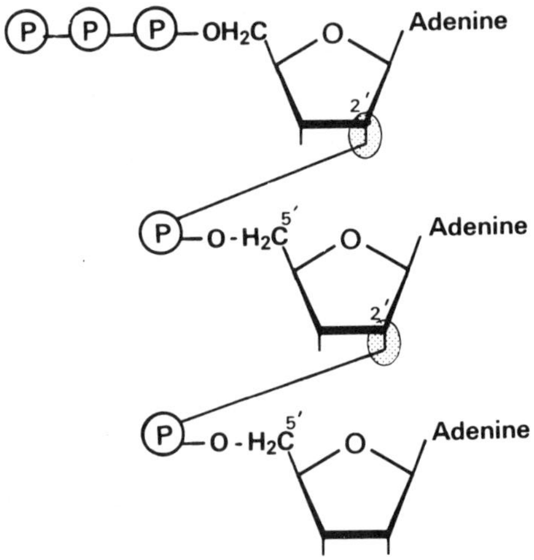

Fig. 122 — Formula of pppA (adenine oligonucleotide (2′–5′)A_n (where n=3,4,5 and even more). Note the 2′–5′ linkage!

(b) Specificity

Interferon acts on different types of virus, but only protects the cells which belong to the same species as the cell which has synthesised it.

The chief consequences of this species specificity is that to protect man from viral infection one must use an interferon obtained not from animal cells, but from human cells.

(c) Different interferons

Different interferons can be obtained:

— alpha interferon, prepared from leukocytes;
— beta interferon, isolated from fibroblasts;
— gamma interferon, obtained from T-lymphocytes.

Until recent years, it was possible to produce only a small amount of interferon in the world, and few therapeutic trials had been performed. However, interferon can now be obtained by genetic engineering and its production is thus increased.

It seems that interferons have not only an antiviral action but may also have an antitumour action in certain cases.

9

Genetic engineering

Genetic engineering is a relatively new scientific field, having been with us for only a few years. The first experiments go back to about 1972, with Berg (who received the Nobel Prize for chemistry in 1980).

The objective is as follows: to manufacture a DNA called 'recombinant DNA', and then to use it for various purposes, in particular to enable cells (for example *E. coli*) to manufacture proteins useful to man.

I. WHAT IS RECOMBINANT DNA?

Recombinant DNA is hybrid DNA obtained *in vitro* by combining two DNAs belonging to different species. For example, a human gene is 'grafted' onto a bacterial DNA.

The techniques consist firstly of isolating the genes involved in this recombination, and then of obtaining the recombinant DNA.

A. Isolation of the eukaryote gene

Several possibilities can be envisaged.

1. Cutting DNA containing the gene required, using a restriction enzyme

Unfortunately, no enzyme exists which cuts the DNA just at the start and the end of a gene. The restriction enzyme will cut at varying distances upstream, downstream, or inside the gene. Furthermore, the restriction enzyme used will also recognise other sites on the DNA and all the DNA will be cut up. A very great number of fragments (100 000 to 1 million) will thus be obtained, containing pieces of genes and non-coding DNA sequnces. This may be viewed as a veritable 'shotgun' aimed at random at the DNA. Next, the gene which is of interest must be selected.

This procedure has a major problem. In effect, the majority of eukaryote genes contain, as previously discussed, coding parts of exons, as well as parts which do not express, or introns. The bacteria into which these genes are introduced do not possess excision–splicing enzymes. Every bacterium, into which a gene with introns and exons is introduced, will transcribe all the gene to give a primary transcript. This primary transcript will be unable to undergo excision–splicing to give mRNA. The whole of the gene with introns and exons will then be translated to finally give an abnormal protein, different from the protein required.

Therefore, procedures must be used which will enable genes without introns to be obtained. Such a gene must be a true complementary copy of the mRNA. THis is called 'cDNA'.

2. Preparation of cDNA

Two procedures allow cDNA to be obtained: chemical synthesis or making a complementary copy of the mRNA.

(a) Chemical synthesis of the gene

The cDNA can be synthesised by assembling the nucleotides which compose it. This chemical synthesis is generally only viable for small genes (e.g. the gene for somatostatin). Recent progress has, however, facilitated obtaining the interferon gene which codes for a protein having more than 100 amino acids. This procedure is not, however, utilised for proteins of a molecular weight of about 60 000 (comprising approximately 600 amino acids, coded by a cDNA sequence of 1800 b.p.).

To use the process of chemical synthesis, the sequence of the gene must obviously be known. In fact, as this procedure is reserved for cases of small genes, it suffices to know the amino acid sequence of the peptide chain coded by this gene. Thanks to the genetic code, a simple deduction allows the nucleotide sequence of the gene to be known. (Nevertheless, given that an amino acid can be coded by several different codons, some uncertainty exists.)

(b) Preparation of cDNA as a complementary copy of mRNA

Firstly the mRNA is isolated from cells where the gene is expressed in relatively large quantities. It is in this way that the mRNAs of globin for example were isolated from erythrocyte stem cells.

The cDNA complementary copy of a mRNA is obtained by means of the reverse transcriptase enzyme. This enzyme (isolated from a retrovirus) is a valuable 'tool' in the genetic engineering laboratory. The model mRNA strand is hydrolysed and the single-stranded cDNA (in the shape of a horseshoe) is converted by cleavage into a double-stranded cDNA to enable it's later insertion into the vector DNA (double-stranded).

B. The DNA vector

The chromosomal DNA of *E. coli* is too long and complicated to be used. The work must be done on a much smaller DNA called a 'vector DNA'. This can be from a plasmid or even a virus (a bacteriophage).

The DNA of the plasmid (or of the bacteriophage) is opened by a restriction enzyme (in general, an enzyme which gives 'adhesive ends' is chosen so as to be able to 'weld' the eukaryote DNA at a later stage). One is, however, limited by the size of the eukaryocyte DNA segment to be grafted.

C. Preparation of recombinant DNA

Here we shall only discuss a few examples which will be simplified as much as possible.

1. Where the gene and the vector DNA have 'adhesive ends'

Adhesive ends are obtained when the restriction enzyme cuts, not in the middle of the palindrome, but around the centre of symmetry. These adhesive ends are complementary when the same restriction enzyme, or certain enzyme couples, have cut the plasmid DNA on the one hand and the human DNA on the other (this is evidently a case where human DNA is prepared according to the 'shotgun' process.) The bringing together of these two DNAs, each having ends formed from a single complementary strand, is easily accomplished by a DNA ligase.

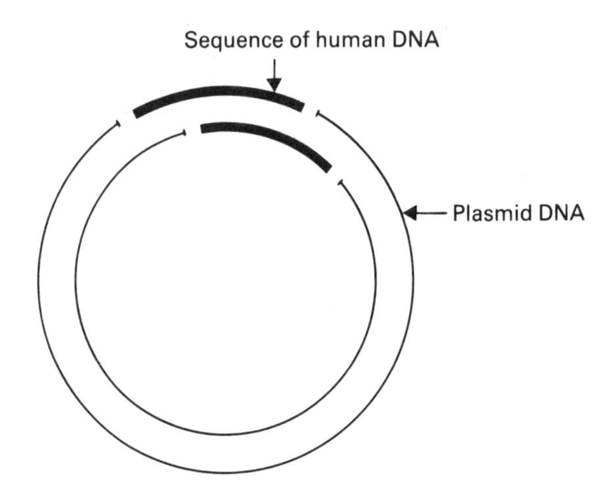

Fig. 123 — Preparation of a recombinant DNA: an example where the ends are adhesive.

2. Where the DNA has 'blunt ends'

Here one can use, for example, another 'tool' also very valuable in a genetic engineering laboratory: 'terminal transferases'. Terminal transferases are enzymes capable of adding a sequence containing only one type of nucleotide to the 3′ end of a chain of nucleotides. A molecule of DNA can also be lengthened dissymetrically, for example with poly-C (poly-C is added to the 3′ end of a DNA chain and also the 3′ end of the other chain of this same DNA). The other molecule of DNA is lengthened in parallel with a poly-G. A DNA ligase then enables the two DNAs to be welded together.

D. Introduction of recombinant DNA into the bacterial cell

The recombinant DNA must then be introduced into *E. coli*. The output is very weak in a case where the vector DNA is a plasmid (attempts can be made to increase this output using techniques which render the bacterial membrane permeable).

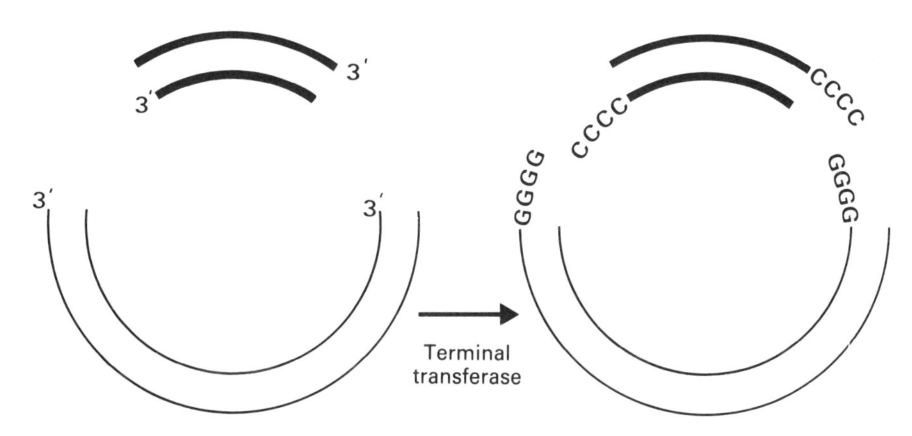

Fig. 124 — Preparation of a recombinant DNA: an example where the ends are blunt. (In this example a terminal transferase has been used to transform the blunt ends into adhesive ends.)

Only a few bacteria will finally contain the recombinant DNA. The selection of these bacteria is therefore necessary.

We shall see later that when expression of the recombinant DNA is required, there is an increasing tendency to use eukaryote cells rather than colibacilli to receive the recombinant DNA.

II. APPLICATIONS

Recombinant DNA is prepared with different objectives:

A. Cloning

1. The aim

The aim of cloning is to isolate and obtain numerous identical copies of a gene. In the bacterium into which the recombinant DNA has been introduced, replication takes place, not only for the DNA of *E. coli*, but also for the recombinant DNA. Thus this procedure provides an extremely useful way to obtain sufficient identical DNA molecules (for example to determine the DNA sequence or to use it for molecular fusion).

2. What is a clone?

A clone is a colony of identical bacteria resulting from the multiplication of a bacterium. In the example used here, a clone is derived from a bacterium which has incorporated a recombinant DNA.

3. cDNA and genome banks

A bank is a collection formed from a great number of clones. cDNA banks are distinct from genome banks.

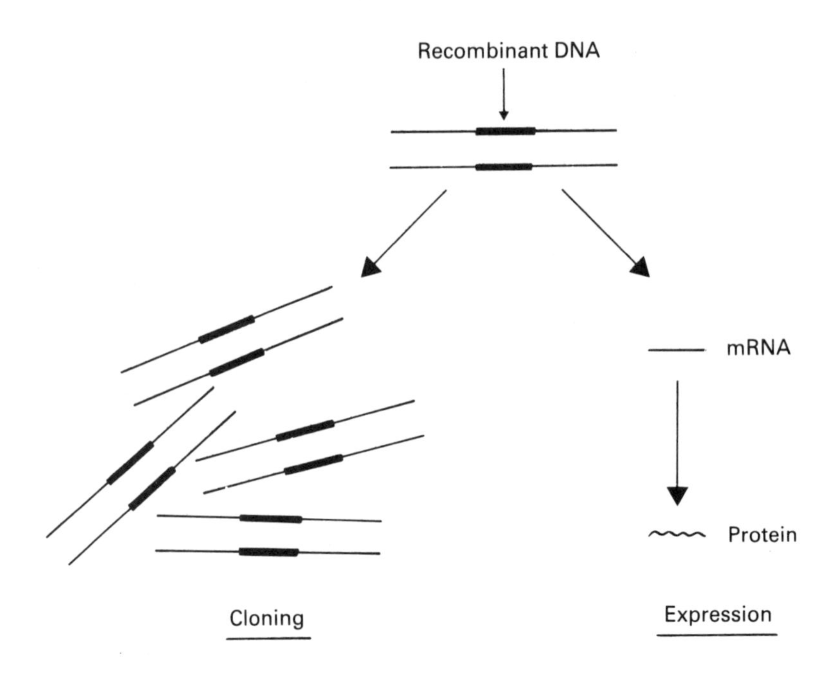

Fig. 125 — Principal applications for recombinant RNA (cloning and expression).

(a) cDNA banks
When the cells in which a desired protein has been synthesised are known, the total mRNAs are separated from these cells. Remember that in a cell not all the genes are expressed. Thus, the isolated mRNAs are only representative of a small percentage of the total genes. The cDNAs are then prepared from these mRNAs. These cDNAs are integrated into vectors which will finally be introduced into bacteria. The different vectors, having incorporated different cDNAs, will constitute a cDNA bank.

(b) Genome banks
When the cells which synthesise a protein are unknown, a theoretically infallible method of finding the gene for the protein is to use a genomic bank. Such a bank is in fact obtained from DNA fragments whose whole represents the totality of the genome. However, it is more difficult to work with a genomic bank than with a cDNA bank as its size is much greater. In fact it contains about 100 times more DNA fragments than a cDNA bank (since it corresponds to all the DNA, coding or non-coding).

4. Screening the banks
Whichever bank is used, it is necessary to screen the bank, i.e. to detect which colony possesses the segment of DNA sought. This operation can be performed with the help of a radioactive probe in a technique called molecular hybridisation (or 'how to search for a needle in a haystack'!)

(a) Molecular hybridisation and probes

• Characteristics of a probe

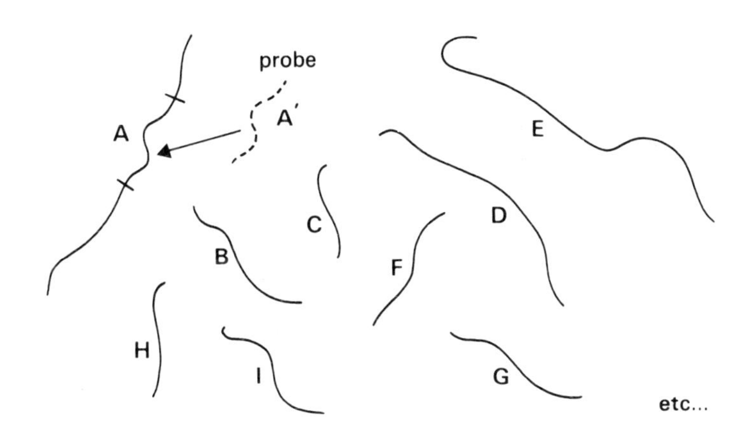

Fig. 126 — Molecular hybridisation: general diagram.

A probe possesses the following characteristics:

— it is a segment of single-stranded nucleic acid (DNA or RNA);

— it is complementary to the segment of nucleic acid to be recognised, this recognition being able to take place DNA–DNA or DNA–RNA (or RNA–RNA). Quite clearly in this hybridisation reaction, the nucleic acids to be recognised must be in the form of one strand. The probe can cover all or part of the nucleic acid segment to be recognised. It is capable of detecting its complementary copy among thousands of different DNA (or RNA) fragments. It is really a hook to catch on to the molecular ladder.

  The diagram below illustrates the differences obtained when one hybridises mRNA with DNA (single stranded) or with the corresponding cDNA (single stranded). In the case of DNA, the introns are not recognised by the mRNA and form loops. On the contrary, with cDNA, hybridisation is perfect.

— The probe should be marked. At present radioactive probes are being used. The positioning of this probe can therefore be easily marked, and consequently the clone to be identified, which will be hybridised with this probe.

• Obtaining a probe

The major difficulty is quite obviously in obtaining a probe. In a case where it has been possible to purify the mRNA, that may be used as a probe. Where this is impossible another process may be used. A small quantity of the protein to be studied is purified and the sequence of a short segment of this protein is determined. The sequence of the segment of the corresponding DNA is then deduced from the genetic code and short samples of DNA are synthesised. However, one can imagine

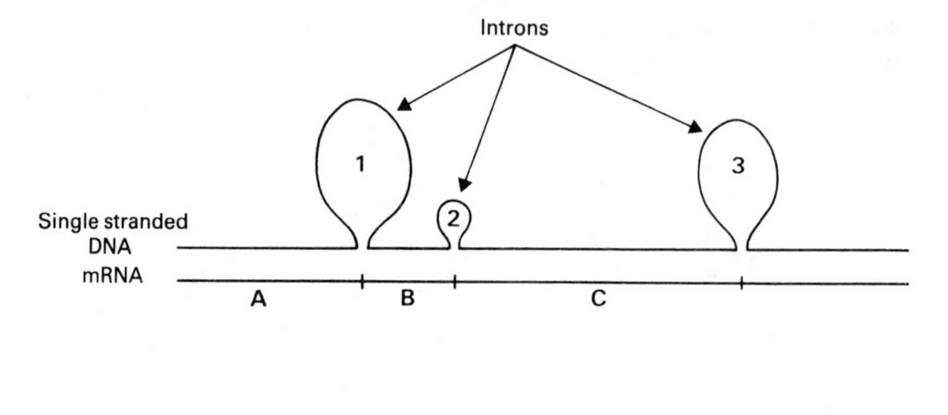

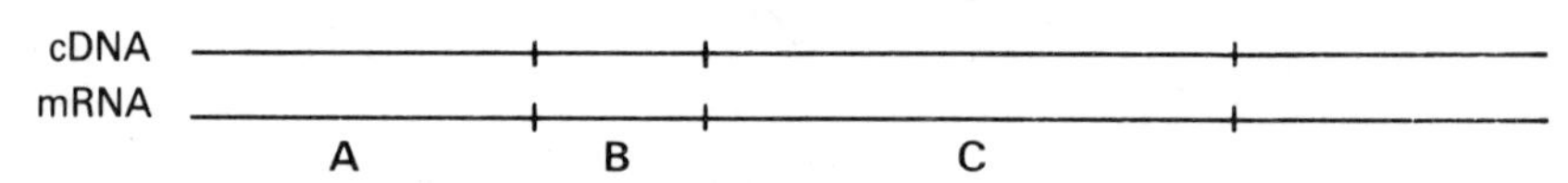

Fig. 127 — Molecular hybridisation between one single-stranded DNA, or cDNA, and the corresponding mRNA.

difficulties encountered when the protein segment sequenced contains amino acids such as Leu, Ser or Arg coded for by six different codons! All these short samples of DNA then have to be synthesised in order to cover all the possibilities.

(b) Immunological detection

Another strategy is possible. For example, if the recombinant DNA is prepared with a vector which allows expression of the gene, the clone synthesising the desired protein could be detected with the aid of a specific antibody.

5. Results

Today many genes are able to be cloned. Amongst others, this procedure has been particularly valuable in obtaining for example the DNA of the hepatitis B virus. We do not know how to culture this virus, however, recombinant DNA containing viral DNA can now be prepared and made to replicate (in mouse fibroblasts).

B. Expressing the gene

This time the aim is to have the eukaryote gene expressed by the bacterium in order to obtain a protein useful to man (e.g. a drug).

1. Difficulties in making a bacterium express a eukaryote gene: the necessity for bacterial signals

The first tests performed were followed by great disappointment: one could obtain a cloning of the integrated gene (which in itself was an extraordinary success) but not an expression of this gene. Thus there was success in obtaining replication of the recombinant DNA but not in transcription. It was feared that the problem of transcription would be an insurmountable barrier. In fact, this difficulty was

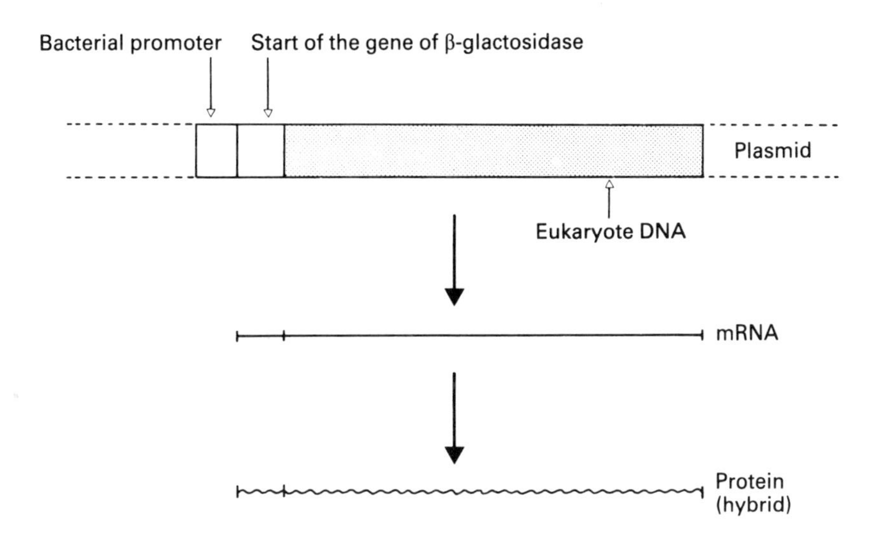

Fig. 128 — Expression of a eukaryote gene by a bacterium: the necessity for bacterial signals.

overcome only a few years later. It sufficed to integrate a signal, recognised by the bacterium, above the eukaryote gene, i.e. a bacterial promoter followed by the start of the bacterial gene. For example, it was possible to use the promoter of the lactose operon and the start of the gene for β-galactosidase.

The bacterial signal allowed the initiation of transcription. The 'deceived' bacterial DNA polymerase transcribes the start of the bacterial gene and then continues to transcribe the human gene. The mRNA formed will give a hybrid protein containing, at the NH_2 end, some supplementary amino acids corresponding to the start of the sequence of the bacterial protein. The protein synthesised must then be extracted, purified and the supplementary amino acids situated at the NH_2 end eliminated. Controls must then be used to verify the absence of toxicity (and in particular the absence of contaminant products of bacterial origin), and to control efficiency.

In certain cases, it is even possible to use only the bacterial signal upstream of ATG and thereby obtain a non-hybrid protein (e.g. interferon).

It may be that the protein obtained does not have the physiological properties expected. In fact the bacterium (as opposed to the eukaryote cells) is not capable of attaching sugars onto a protein, and if the protein studied is a glycoprotein, it can be inactive if it only comprises the peptide chain.

At present, many proteins have been synthesised in this way, for example insulin, interferon, certain coagulation factors, hormones, vaccines, etc. However, before the tests for tolerance, activity, etc., are all carried out, several more years will have to elapse. The first commercialised protein obtained by genetic engineering techniques was insulin.

To sum up, if it is desired that an animal gene is expressed in a bacterium, it is necessary to integrate not only a gene without introns, but also bacterial signals upstream of this gene.

2. *Use of mammalian cells in place of* E. coli

There is an increasing tendency to use mammalian cells rather than bacteria to express recombinant DNA. The human growth hormone has thus been synthesised (in 1983) in cultured monkey cells. In 1984 clotting factor VIII was synthesised in the kidney cells of the monkey and the hamster. Other types of eukaryote cells continue to be used.

3. *Some examples of protein syntheses*

(a) Insulin

As we have already seen in the section 'Signal sequence', insulin is a hormone synthesised by the pancreas and which is lacking in certain types of diabetes. In these patients, exogenous insulin must be administered.

- Recap of the formula for animal insulins

Remember that insulin is formed from two, peptide chains — A(21 amino acids) and B(30 amino acids) — connected by two disulphide bridges (a third intra-chain A disulphide bridge, also exists).

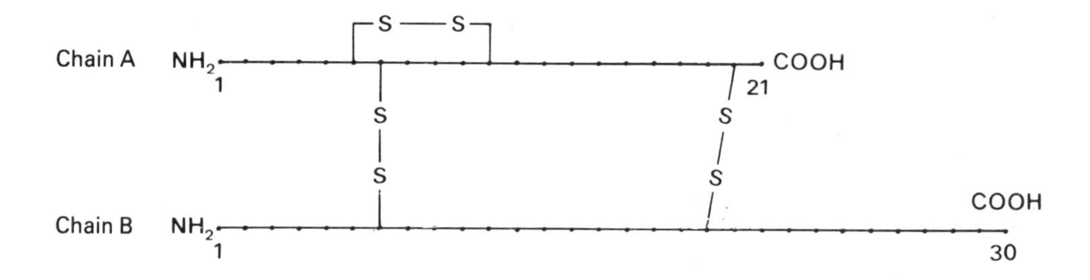

Fig. 129 — Diagram representing the structure for insulin.

Bovine and porcine insulins have a structure very close to human insulin. Thus, bovine insulin differs by one amino acid: no. 30 situated at the COOH terminal of the B chain. These insulins can be used pharmacologically. They are obtained by extraction from bull or pig pancreas material obtained from abbatoirs. They must be carefully purified to eliminate as many antigenic contaminants as possible.

- Obtaining human insulin

In certain cases insulin of animal origin cannot be used, (bovine or porcine obtained by extraction) and obtaining human insulin by chemical synthesis is too onerous. At present two processes can be used to produce human insulin.

— Semi-synthetic insulin.

Porcine insulin, as detailed above, only differs from human insulin by the amino acid situated at the COOH end of the B chain (Ala in the pig, Thr in man). To obtain insulin identical to human insulin it is possible simply to change the last amino acid of the B peptide chain (by a reaction called transpeptidation). This replacement of Ala

Type of insulin	Chain A		Chain B
	aa 8	aa 10	aa 30
BOVINE	Ala	Val	Ala
PORCINE	Thr	Ile	Ala
HUMAN	Thr	Ile	Thr

Fig. 130 — Comparison of bovine, porcine and human insulin formulae. Human insulin differs from: — porcine insulin by a single amino acid (no. 30 of chain B), — bovine insulin by three amino acids (no. 30 of chain B, no. 8 and 10 of chain A).

(porcine) by Thr occurs in the course of an enzymatic stage where trypsin intervenes, according to an original process developed by the manufacturers of this semi-synthetic insulin.

— human insulin obtained by genetic engineering techniques.
The A and B chains of insulin are not coded for by two distinct genes. One must remember that insulin is synthesised in the form of a polypeptide precursor formed from a single chain. Thus only the gene coding for preproinsulin exists.

The laboratories producing insulin by genetic engineering have chosen a fabrication process which consists firstly of chemically synthesising the genes which correspond to the A and B chains. Various operations must then be performed, briefly summed up as follows:

— fusion of the gene corresponding to the A chain with a bacterial promoter and a bacterial gene,
— formation of recombinant DNA using a plasmid,
— incorporation of this plasmid in *E. coli*,
— recovery of the peptide A chain and purification,
— performance of the above operations to obtain the peptide B chain,
— assembly of the peptide A chain with the peptide B chain, which is not without problems, since the NH_2–COOH polarities of the two chains forming insulin must be respected.

(b) Vaccine against hepatitis B virus
The principle of all types of vaccination is to inject an antigen into the subject to be protected. Antibodies will then be synthesised in response to this antigen.

Remember that the hepatitis B virus contains DNA as its nucleic acid protected by a protein capsule which is itself surrounded by a protein envelope. The antigen used in the preparation of the hepatitis B vaccine is the protein part corresponding to hepatitis B antigens. The first stage in obtaining a vaccine against hepatitis B is thus the isolation of these proteins.

- Present-day means

Healthy carrier subjects are chosen (who have previously been parasitised by the hepatitis B virus). The blood of these subjects no longer contains any virus although empty envelopes are found there (without DNA). These empty envelopes can be extracted from the blood of these subjects, theoretically without risk of them being contaminated by viral DNA.

- Future means

It is probable that the number of healthy carrier subjects containing empty envelopes in their blood will diminish if vaccination is intensified. Therefore a method must be found which makes the process of preparation easier. Genetic engineering techniques have now been used for the preparation of HB antigen in which recombinant DNA is introduced into a cultured eukaryote cell (e.g. mouse fibroblast) *in vitro*.

(c) Factor VIII

Factor VIII is a protein which plays an important role in coagulation. The absence of functional factor VIII is responsible for haemophilia A. As the gene coding for this protein is carried by the sex chromosome X, haemophilia is transmitted by women and affects male infants. (Thus Queen Victoria of England transmitted haemophilia to about ten of her male descendents, one of whom was the Tzarovitch Alexis of Russia.) A woman can in fact possess one abnormal gene without being haemophilic since the normal gene carried by the other X chromosome protects her. As it is extremely unlikely that both genes would be abnormal, haemophilia in women is very rare. In males who have two different sex chromosomes, X and Y, the Y chromosome is incapable of compensating for an abnormality of the single X chromosome.

At present, haemophilic subjects are treated by injections of factor VIII isolated from donated blood. It is hoped that the production of factor VIII by genetic engineering will enable a purer product to be obtained with less difficulty.

The gene of factor VIII is now being cloned and sequenced. Four publications in the same issue of the journal *Nature* (November 1984) report on this work.

The factor VIII gene is very long. It contains 186 000 b.p. with 26 exons! The factor VIII obtained from the corresponding cDNA is a protein of 2332 amino acids (2351 amino acids minus 19 amino acids for the signal sequence).

(d) Factor IX

Factor IX is another protein which plays an important role in blood coagulation. It is lacking in people with haemophilia B. This protein contains 415 amino acids and is glycosylated. The gamma-carboxylation of 12 Glu residues is essential to its function, and *in vivo* this process is dependent on vitamin K. This active form of

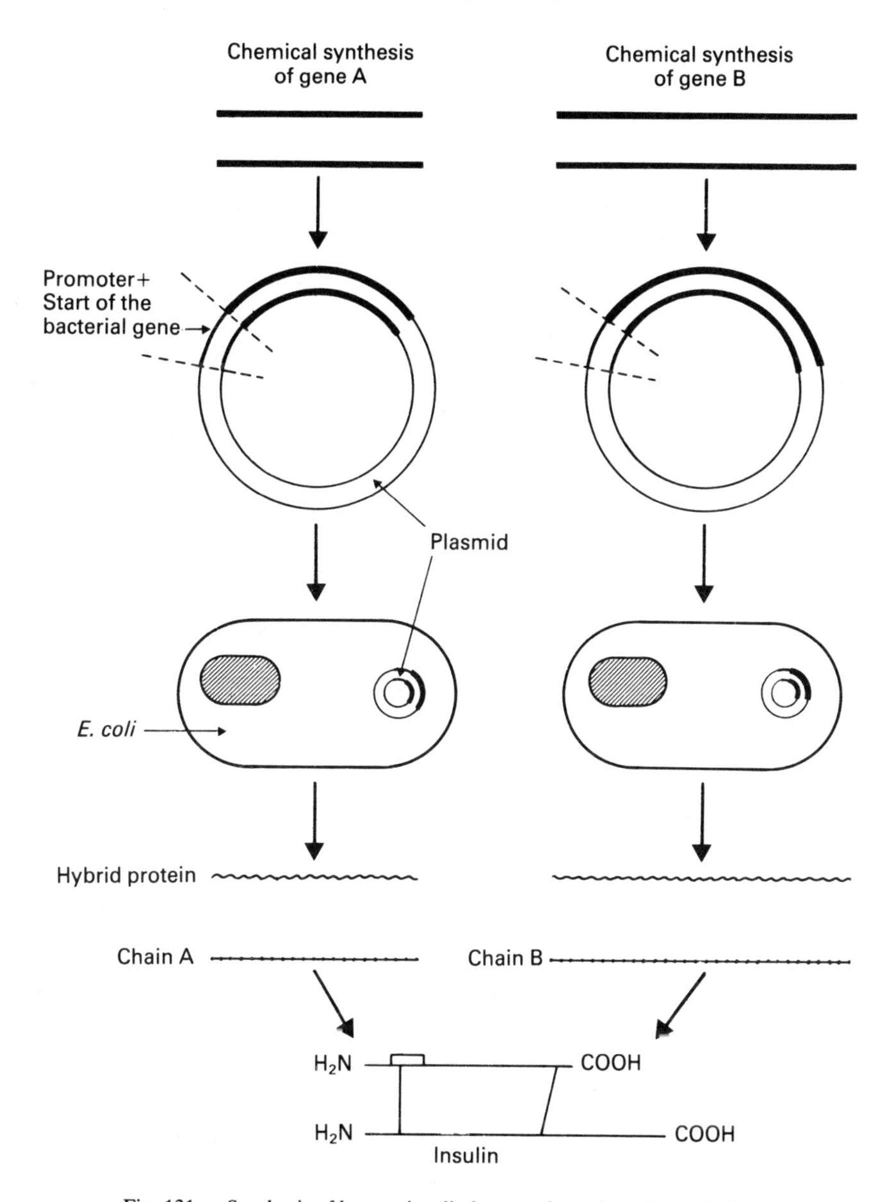

Fig. 131 — Synthesis of human insulin by genetic engineering techniques.

factor IX has been obtained by genetic engineering techniques (in 1985) on cultured cells (mainly with progeny of hepatoma cells — the liver being the principal tissue capable of carrying out the carboxylation process — but also with progeny of kidney cells).

(e) The von Willebrand factor

Cloning of the gene coding for the von Willebrand factor (which also plays an important role in coagulation) has been achieved recently from cultured endothelial cells (the gene for this clotting factor is localised on chromosome 12).

Clearly these are only a few examples chosen at random from the many results of the last few years.

C. Microinjection of genes into embryos

A new type of experiment has been started with the aim of manipulating embryos genetically. These experiments consist of introducing a gene of foreign origin into the nucleus of a newly fertilised egg, then reimplanting the altered egg into the uterus of a female carrier. An individual which has incorporated the foreign gene will eventually be born (the efficiency of the operation is not good). This type of experiment encounters many obstacles: in particular the inability, in the present state of our technques, to integrate a gene into the position it normally occupies on the chromosome (grafted genes seem to be incorporated haphazardly, and not necessarily in a site which will allow expression). Nevertheless, surprising results have been obtained. It has been possible to inject the growth hormone of the rat into a mouse embryo. Giant mice the size of rats were produced in this way. However, to obtain mice as big as rats is obviously not of great practical interest; the hope is that one day it will be possible to graft genes into human embryos and thus manage to correct a specific genetic defect. We are still very far, however, from being able to incorporate a functional gene into a human embryo, and into a precise spot on the genome. In addition it must be possible to regulate the expression of the gene introduced: the presence of a protein in excess can be as dangerous as the absence of this protein.

D. Examples of diagnoses of hereditary diseases by genetic engineering techniques using molecular hybridisation

Certain diagnoses are now possible using genetic engineering techniques. These techniques call upon 'molecular hybridisation'.

1. Research into hepatitis B virus

The biological diagnosis of the hepatitis B virus is usually achieved indirectly by investigating the blood for antigen markers (HBs, HBe) and/or by detection of anti-HBs, anti-HBc, anti-HBe antibodies.

It is now possible, in certain cases, to detect the DNA of the virus directly by molecular hybridisation. These techniques are at present reserved for very specialised laboratories (e.g. The Pasteur Institute).

(a) In the blood

The patients's serum is treated in order to extract the viral DNA.

- This DNA is denatured (obtaining single-stranded DNA) and then fixed on a nitrocellulose filter (a type of filter paper).
- Next, molecular hybridisation is undertaken using radiolabelled cloned DNA (similarly denatured) of the hepatitis B virus as a probe. The filter is then washed to eliminate the excess of the non-hybridised probe.

— The molecular hybridisation is revealed by autoradiography: a spot is observed if the serum contains the hepatitis B virus. This method is much more sensitive than the classic techniques.

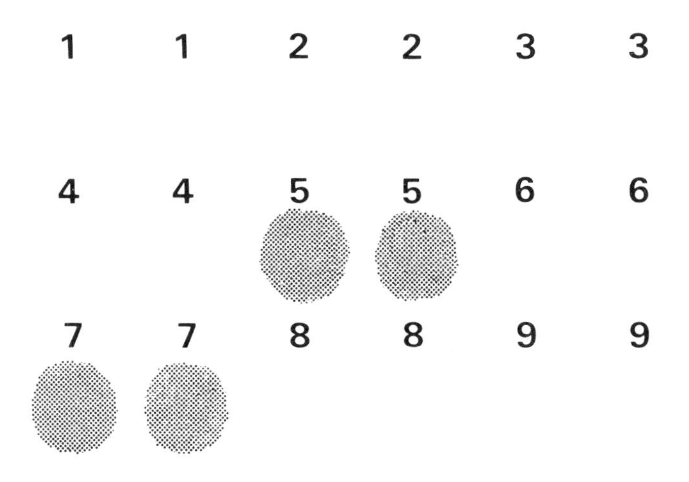

Fig. 132 — Testing the serum of nine subjects with hepatitis B virus by genetic engineering techniques (diagram representing an autoradiogram).

(b) In hepatic tissue
It is equally possible to search for the DNA of the hepatitis B virus in hepatocytes obtained from biopsies. It is even possible to determine of the viral DNA is free or integrated into the DNA of human hepatocytes.

2. Prenatal diagnosis
Only the example of the prenatal diagnosis of sickle-cell anaemia will be given here.

(a) The mutation responsible for the abnormality
In the chapter on mutations we saw that this disease is due to a simple mutation at the level of the sixth codon (first exon) of the gene of the β-globin chain where T replaces A (which is translated into an abnormal protein having Val instead of Glu in the sixth position). Haemoglobin having such an abnormality is known as haemoglobin S.

There is no known means of treating sickle-cell anaemia. The only possibility presently offered to families at risk is to carry out prenatal diagnosis for the condition. If the result is positive and obtained sufficiently early, a therapeutic abortion can then be contemplated by the parents.

Until recent years, however, it was not possible to make such an early diagnosis. It was only possible to investigate, towards the fourth month of pregnancy, the presence of an abnormal haemoglobin in the fetal blood *in utero*. The technique used to diagnose abnormal haemoglobin is electrophoresis. In fact, the electrophoretic migration of haemoglobin varies depending on whether the amino acid in position 6 is neutral (Val) or charged (Glu).

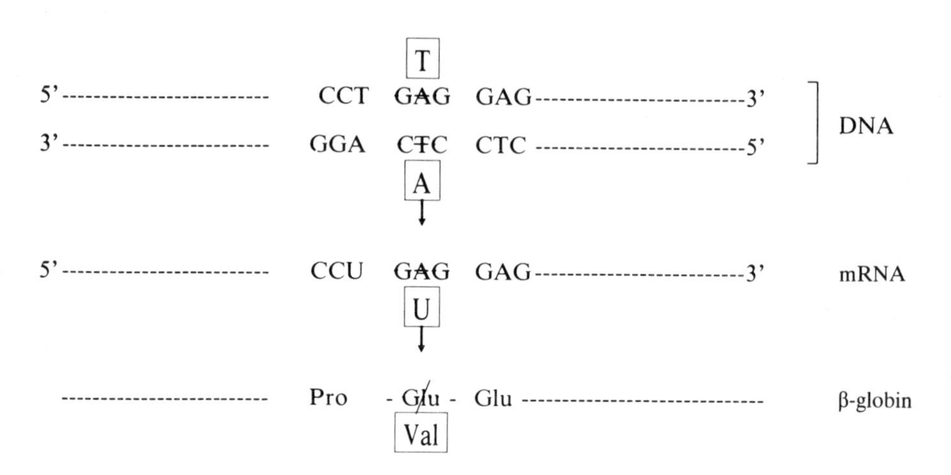

Fig. 133 — Sickle-cell anaemia: abnormalities observed at the levels of the gene, of the mRNA and of the protein expressed (β-globin).

At present it is possible, in specialised laboratories, to directly study the abnormality at the gene level (instead of, as in the previous case, investigating at the level of the protein — the product of gene expression). This test is performed on cells belonging to the fetus; cells of amniotic fluid sampled from the 17th week of pregnancy, or even earlier (8th week) on cells from the trophoblast (superficial layer of placental villus cells) sampled via the vagina.

(b) Principle of the diagnostic technique

To detect mutations at the gene of the sixth codon, a particular restriction enzyme is used, MstII, (extracted from the blue alga 'microcoleus'). This enzyme recognises the palindrome:

C C|T N A G G
G G A N'T|C C

where N can be any of the four bases (as it happens this is an imperfect palindrome). It is noticeable that this palindrome corresponds exactly to the sequence of codons 5, 6, and the start of 7, situated in exon 1 of the gene of the β-globin (site y).

... CCT GAG G AG...
... GGA CTC C TC...

5th 6th 7th

Just before the second exon, at the end of the first intron, there is one site (z) recognised by the enzyme MstII. The same applies to the zone preceding the globin gene which is called the 'flanking region' (site x). The MstII restriction enzyme then cuts the globin gene at these three points, x,y,z, thus producing two fragments of 1.15 and 0.20 kb. (kilobases) respectively. On the other hand, if there is a mutation at the sixth codon (site y) the MstII enzyme no longer recognises the central sequence and will no longer cut at y. A single fragment of 1.35 kb. will be obtained (the cuttings at x and z continue to occur since no mutation is produced here).

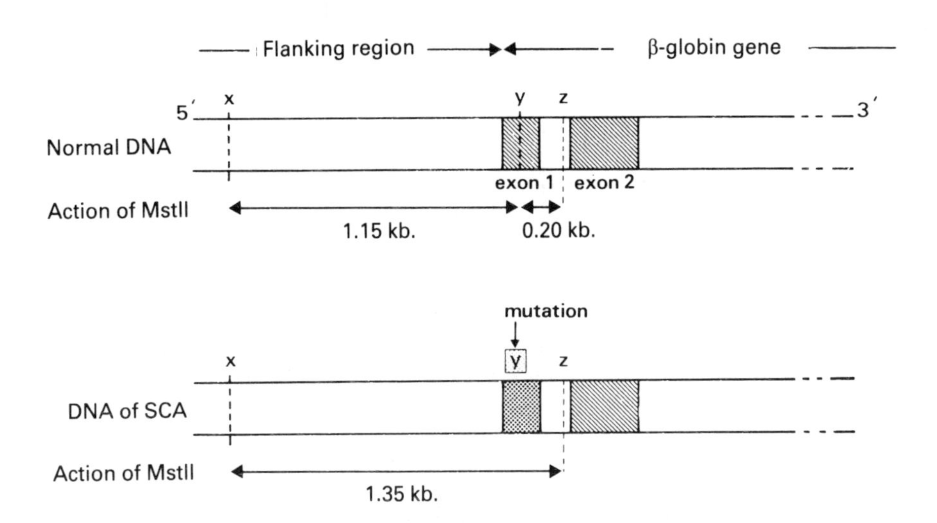

Fig. 134 — Diagnosis of sickle-cell anaemia (SCA) by genetic engineering techniques: the restriction enzyme MstII no longer recognises the site situated in exon 1 of the β-globin gene. A fragment of 1.35 kb. (instead of 1.15 kb.) is produced.

Thus the techniques of diagnosis by molecular hybridisation consists of carrying out the following operations:

— Extraction of cDNA from appropriate cells (amniotic fluid fibroblasts or trophoblastic tissue).
— Cutting of DNA by the restriction enzyme MstII, giving fragments of different lengths.
— Separation of these fragments according to length by gel electrophoresis.
— Denaturation of the DNA (separation of the two strands of the double helix).
— Transfer onto a nitrocellulose filter.
— Molecular hybridisation with a specific radioactive probe. This probe is prepared, for example, with the 1.15 kb. fragment obtained after cutting at x and y of the cloned gene of the normal β-globin. It will hybridise equally well with a 1.15 kb.

fragment (in the case of normal haemoglobin A) as with a 1.35 kb. fragment (in the case of haemoglobin S).

— Demonstration by autoradiography. The autoradiogram shows three possibilities, depending on whether the subject is normal (a spot at 1.15), possesses haemoglobin S in the homozygotic state (a spot at 1.35) or in the heterozygotic state (two spots, at 1.15 and 1.35).

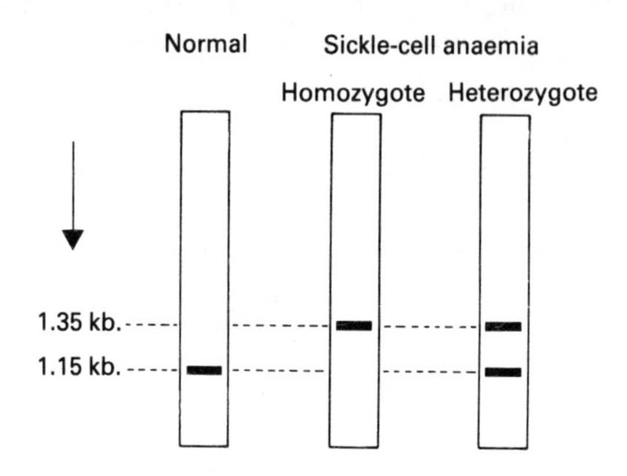

Fig. 135 — Diagram of autoradiograms in the diagnosis of sickle-cell anaemia by genetic engineering techniques using the restriction enzyme MstII.

These are very recent techniques which can only be applied in very specialised laboratories.

This approach to prenatal diagnosis using molecular probes is now beginning to be used for other hereditary diseases. However, not all the genetic diseases are as well defined as sickle-cell anaemia, and not all are due to a single mutation in the proximity of a single gene.

One can only enthuse about the considerable progress of genetic engineering techniques which offer us new diagnostic tools and which will increasingly influence the practice of medicine. About 200 abnormal genes responsible for human diseases, have been localised on the chromosomes. Today, medical students who learn macroscopic anatomy will soon have to learn another type of anatomy.

10

Selected bibliography

DNA

Structure of DNA

— The helical repeat of B-DNA in solution, in physiological conditions is 10.3–10.5 b.p. per turn:
  Wang, J. C. Helical repeat of DNA in solution. *Proc. Natl. Acad. Sci., USA* 1979 **76** 200–203.

Topoisomerases

— General reviews on topoisomerases:
  Gellert, M. DNA topoisomerases. *Ann. Rev. Biochem.* 1981 **50** 879–910.
  Wang, J. C. DNA topoisomerases. *Ann. Rev. Biochem.* 1985 **54** 665–697.

— New topiosomerases. At present only examples of topoisomerases responsible for negative superturns are known. An enzymatic fraction has recently been discovered (in archeobacteria) capable of giving a positive DNA superturn:
  Kikuchi, A. and Asai, K. Reverse gyrase, a topoisomerase which introduces positive superhelical turns into DNA. *Nature*, 1984 **309** 677–681.

Z-DNA

— General review:
  Rich, A., Nordheim, A. and Wang, A. H. J. The chemistry and biology of left-handed Z-DNA. *Ann. Rev. Biochem.* 1984 **53** 791–846.

— Preparation of anti-Z-DNA antibodies used to detect the presence of Z-DNA in biological systems:
  Lafer, E. M., Moller, A., Nordheim, A., Stollar, B. D. and Rich, A. Antibodies specific for left-handed Z-DNA. *Proc. Natl. Acad. Sci. USA* 1981 **78** 3546–3550.
  Viegas–Pequignot, E., Derbin, C., Malfoy, B., Taillandier, E., Leng, M., and Dutrillaux, B. Z-DNA immunoreactivity in fixed metaphase chromosomes of primitives. *Proc. Natl. Acad. Sci. USA* 1983 **80** 5890–5894.

— Detection of anti-Z-DNA antibodies:

Lafer, E. M., Valle, R. P. C., Moller, A., Nordheim, A., Shur, P. H., Rich, A. and Stollar, B. D. Z-DNA specific antibodies in human systemic lupus erythematosus. *J. Clin. Invest.* 1983 **71** 314–321.

Restriction enzymes

— List of the chief restriction enzymes and sequences of the sites they recognise:

Roberts, R. J. Restriction endonucleases. In Linn, S. M. and Roberts, R. J. (eds) *Nucleases*, Cold Spring Harbor Laboratory, New York, 1982, pp. 312–340.

— Endonucleases exist which cut a sequence of DNA elsewhere than in a palindrome:

Yuan, R. Structure and mechanism of multifunctional restriction endonucleases. *Ann. Rev. Biochem.* 1981 **50** 285–315.

Determination of DNA sequences

— Maxam and Gilbert's method:

Maxam, A. M. and Gilbert, W. A new method for sequencing DNA. *Proc. Natl. Acad. Sci. USA.* 1977 **74** 560–564.

— Sanger's method:

Sanger, F. and Coulson, A. R. A rapid method for determining sequences in DNA by primed synthesis with DNA polymerase. *J. Mol. Biol.* 1975 **94** 441–448.

Sanger, F., Coulson, A. R., Barrell, B. G., Smith, A. J. H. and Roe, B. A. Cloning in single-stranded bacteriophage as an aid to rapid DNA sequencing. *J. Mol. Biol.* 1980 **143** 161–178.

TRANSCRIPTION

The promoter

— In prokaryotes:

McClure, W. R. Mechanisms and control of transcription initiation in prokaryotes. *Ann. Rev. Biochem.* 1985 **54** 171–204.

Manley, J. L. Analysis of the expression of genes encoding animal mRNA by *in vitro* techniques. *Progr. Nucleic Acid Res. Mol. Biol.* 1983 **30** 196–244.

Transcription in eukaryotes

— General review:

Nevins, J. R. The pathway of eukaryotic mRNA formation. *Ann. Rev. Biochem.* 1983 **52** 441–466.

— Introns:

Definition of introns and exons by Gilbert (Nobel Prize):

Gilbert, W. Why genes in pieces? *Nature* 1978 **271** 501.

An example of a discontinued gene described by the French team of Chambon, the gene coding for ovalbumin contains seven exons separated by six introns:

Gannon, F., O'Hare, K., Perrin, F., Le Pennec, J. P., Benoist, C., Cochet, M., Breathnach, R., Royal, A., Garapin, A., Cami, B. and Chambon, P.

Organisation and sequences at the 5′ end of a cloned complete ovalbumin gene. *Nature* 1979 **278** 428–434.

Transcriptions of introns are excised, giving 'lasso' (or lariat) shapes:

Keller, W. The RNA lariat, a new ring to the splicing of mRNA precursors. *Cell* 1984 **39** 423–425.

The snRNAs have a complementary part (towards the 5′ end) of the intron transcription:

Pettersson, I., Hinterberger, M., Mimori, T., Gottlieb, E. and Steitz, J. A., The structure of mammalian small nuclear ribonucleoproteins. *J. Biol. Chem.* 1984 **259** 5907–5914.

Anti sn-RNA antibodies were found in lupus (in fact snRNAs are not themselves antigenic, but are associated with proteins which confer this antigenicity on them).

Lerner, M. R. and Steitz, J. A. Antibodies to small nuclear RNAs complexed with proteins are produced by patients with systemic lupus erythematosus. *Proc. Natl. Acad. Sci. USA* 1979 **76** 5495–5499.

Cross-splicing: until recently only *cis* splicing (intramolecular) has been known, i.e. the transcription of spliced exons belonging to the same primary transcript. For a short while cross-splicing (intramolecular) has been known of in two exon transcripts belonging to different primary transcripts:

Solnick, D., Trans-splicing of mRNA precursors. *Cell* 1985 **42** 157–164.

Konarska, M. M., Padgett, R. A. and Sharp, P. A. Trans splicing of mRNA precursors. *Cell* 1985 **42** 165–171.

Introns coding for a maturase (Slonimski's team):

Jacq, C., Baroques, J., Becam, A. M., Slonimski, P. P., Guiso, N. and Danchin, A. Antibodies against a fused 'lacZ-yeast mitochondrial intron' gene product allow identification of the mRNA maturase coded by the fourth intron of the yeast cob-box gene. *Embo J.* 1984 **3** 1562–1572.

Guiso, N., Dreyfus, M., Siffert, O., Danchin, A., Spyridakis, A., Gargouri, A., Claisse, M. and Slonimski, P. P. Antibodies against synthetic oligopeptides allow indentification of the mRNA-maturase encoded by the second intron of the yeast cob-box gene. *Embo J.* 1984 **3** 1769–1772.

Introns were not found only in eukaryotes. They were also found in Archeobacteria:

Kaine, B. P., Gupta, R. and Woese, C. R. Putative introns in tRNA genes of prokaryotes. *Proc. Natl. Acad. Sci. USA*. 1983 **80** 3309–3312.

Supergene families and mosaic proteins (the concept of Hood *et al.*) developed by Brown, Goldstein *et al.* (Nobel Prize):

Sudhof, T. C., Goldstein, J. L., Brown, M. S. *et al. Science* 1985 **228** 815–822, 893–895.

Hood, L., Kronenberg, M. and Hunkapiller, T. T-cell antigen receptors and the immunoglobulin supergene family. *Cell* 1985 **40** 225–229.

— The cap:

The 'cap' is necessary for translation in eukaryotes:

Paterson, B. M. and Rosenberg, M. Efficient translation of prokaryotic mRNA in a eukaryotic cell-free system requires addition of a cap structure. *Nature* 1979 **279** 692–696.

— Poly-A:

Littauer, U. Z. and Soreq, H. The regulatory function of poly (A) and a adjacent 3′ sequences in translated RNA. *Prog. Nucleic Acid Res. Mol. Biol.* 1982 **27** 53–83.

'Enhancers' and 'silencers'

— 'Enhancers':

Gillies, S. D., Morrison, S. L., Oi, V. T. and Tonegawa, S. A tissue-specific transcription enhancer element is located in the major intron of a rearranged immunoglobulin heavy chain gene. *Cell* 1983 **33** 717–728.

— 'Silencers':

Brand, A., Breeden, L., Abraham, J., Sternglanz, R. and Nasmyth, K. Characterisation of a 'silencer' in yeast: a DNA sequence with properties opposite to those of a transcriptional enhancer. *Cell* 1985 **41** 41–48.

TRANSLATION

The genetic code

— Overlap in the code was detected during determination of the sequence of viruses such as:

- $\phi\times174$ (bacteriophage)

  Barrel, B. G., Air, G. M. and Hutchinson, C. A. Overlapping genes in bacteriophage $\phi\times174$, *Nature* 1976 **264** 34–41.

  Sanger, F., Air, G. M., Barrell, B. G., Brown, N. L., Coulson, A. R., Fiddes, J. C., Hutchinson, C. A., Slocombe, P. M. and Smith, M. Nucleotide sequence of bacteriophage $\phi\times174$ DNA. *Nature* 1977 **265** 687–695.

- SV40 (Simian virus)

  Reddy, V. B., Thimmappaya, B., Dhar, R., Subramanian, K. N., Zain, B. S., Pan, J., Ghosh, P. K., Celma, M. L. and Weissman, S. M. The genome of simian virus 40. *Science* 1978 **200** 494–502.

  Fiers, W., Contreras, R., Haegeman, G., Rogiers, R., Van de Voorde, A., Van Heuverswyn, H., Van Herreweghe, J., Volckaert, G. and Ysebaert, M. Complete nucleotide sequence of SV40 DNA. *Nature* 1978 **273** 113–120.

— Exceptions to the universal code:

- in human mitochondria

  Anderson, S., Bankier, A. T., Barrell, B. G., de Bruijn, M. H. L., Coulson, A. R., Drouin, J., Eperon, I. C., Nierlich, D. P., Roe, B. A., Sanger, F., Schreier, P. H., Smith, A. J. H., Staden, R. and Young, I. G. Sequence and organization of the human mitochondrial genome. *Nature* 1981 **290** 457–465.

- in *Paramecium*

Preer, J. R., Preer, L. B., Rudman, B. M. and Barnett, A. J. Deviation from the universal code shown by the gene for surface protein 51A in *Paramecium*. *Nature* 1985 **314** 188–190.

Wobble

— Some characteristics of mitochondria:

Discovery of the U/N wobble by Sanger and his team:

Barrell, B. G., Anderson, S., Bankier, A. T., de Bruijn, M. H. L., Chen, E., Coulson, A. R., Drouin, J., Eperon, I. C., Nierlich, D. P., Roe, B. A., Sanger, F., Scheier, P. H., Smith, A. J. H., Staden, R. and Young, I. G. Different pattern of codon recognition by mammalian mitochondrial tRNAs *Proc. Natl. Acad. Sci. USA* 1980 **77** 3164–3166.

— 22 tRNAs identified (instead of 32 in the cytoplasmic system):
Clayton, D. A. Transcription of the mammalian mitochondrial genome. *Ann. Rev. Biochem.* 1984 **53** 573–594.

Initiation, elongation and termination factors

— General review:
Moldave, K. Eukaryotic protein synthesis. *Ann. Rev. Biochem.* 1985 **54** 1109–1149.

The signal sequence and SRP (signal recognition particle)

Walter, P. and Blobel, G. Signal recognition particle contains a 7S RNA essential for protein translocation across the endoplasmic reticulum. *Nature* 1982 **299** 691–698.

Meyer, D. I., Louvard, D. and Dobberstain, B. Characterization of molecules involved in protein translocation using a specific antibody. *J. Cell Biol.* 1982 **92** 579–583.

Glycosylation of proteins

Kornfeld, R. and Kornfeld, S. Assembly of asparagine-linked oligosaccharides. *Ann. Rev. Biochem.* 1985 **54** 631–664.

Synthesis of immunoglobulins:

— Altering of immunoglobulin genes. (Note: differentiate carefully the results obtained in man from those obtained in the mouse).
Leder, P. Genetic control of immunoglobulin production. *Hosp. Pract.* 1983 **18** 73–82.
Honjo, T. Origin of immune diversity: genetic variation and selection. *Ann. Rev. Biochem.* 1985 **54** 803–830.

REGULATION OF THE SYNTHESIS OF PROTEINS AND EXPRESSION OF GENES

Example of the regulation at translation level with ribosomal proteins

— General review on the structures and function of the ribosome:
Lake J. A. Evolving ribosome structure: domains in archaebacteria, eubacteria, eocytes and eukaryotes. *Ann. Rev. Bioch.* 1985 **54** 507–530.

— Regulation of the synthesis of ribosomal proteins:
Nomura, M., Gourse, R. and Baughman, G. Regulation of the synthesis of ribosomes and ribosomal components. *Ann. Rev. Biochem.* 1984 **53** 75–117.

Chromatin and nucleosomes

Mathis, D., Oudet, P. and Chambon, P. Structure of transcribing chromatin. *Prog. Nucleic Acid Res. Mol. Biol.* 1980, **24** 1–55.

Hypomethylation of genes and expression

— Relations between hypomethylation of DNA and expression of genes, maintenance methylases, relations between hypomethylation and cancer, etc.:

Riggs, A. D. 5-methylcytosine, gene regulation and cancer. *Adv. Cancer Res.* 1983 **40** 1–30.

Doerfler, W. DNA methylation and gene activity. *Ann. Rev. Biochem.* 1983 **52** 93–124.

Feinberg, A. P. and Vogelstein, B. Hypomethylation distinguishes genes of some human cancers from their normal counterparts. *Nature* 1983 **301** 89–92.

— Certain restriction enzymes permit recognition of whether or not DNA sequences are methylated:

Mandel, J. L. and Chambon, P. DNA methylation: organ specific variations in the methylation pattern within and around ovalbumin and other chicken genes. *Nucleic Acid Res.* 1979 **7** 2081–2103.

Queuine

— Queuine, an atypical base encountered in certain tRNAs could play a role in cellular differentiation and cancer:

Nishimura, S. Structure, biosynthesis and function of queuosine in transfer RNA. *Prog. Nucleic Acid Res. Mol. Biol.* 1983 **28** 49–73.

Embryonic proteins and proteins called 'thermal shock' proteins

Bensaude, O., Babinet, C., Morange, M. and Jacob, F. Heat shock proteins first major products of zygotic gene activity in mouse embryo. *Nature* 1983 **305** 331–333.

Study of genes which control the early stages of embryonic development

Gehring, W. The homeo box: a key to the understanding of development?. *Cell* 1985 **40** 3–5.

Hormones

— The new second messengers, IP3 and DG:

Hokin, L. E. Receptors and phosphoinositide-generated second messengers. *Ann. Rev. Biochem.* 1985 **54** 205–235.

REPLICATION

Replication in prokaryotes

Nossal, N. G. Prokaryotic DNA replication systems. *Ann. Rev. Biochem.* 1983 **53** 581–615.

Replication in eukaryotes

Challberg, M. D. and Kellt, T. J. Eukaryotic DNA replication: viral and plasmid model system. *Ann. Rev. Biochem.* 1982 **51** 901–934.

'Editor function' enzymes

Loeb, L. A. and Kunkel, T. A. Fidelity of DNA synthesis. *Ann. Rev. Biochem.* 1982 **51** 429–457.

REPAIR OF DNA

Repair systems of DNA

Walker, G. C. Inducible DNA repair systems. *Ann. Rev. Biochem.* 1985 **54** 425–457.

VIRUSES

Retroviruses

- LTR sequences:
  Chen, H. R. and Barker, W. C. Nucleotide sequences of the retroviral long terminal repeats and their adjacent regions. *Nucleic Acids Res.* 1984 **12** 1767–1778.

The AIDS virus

— Determination of the sequence by:
- The French team:
  Wain-Hobson, S., Sonigo, P. Dangos, O., Cole, S. and Alizon, M. Nucleotide sequence of the AIDS virus LAV. *Cell* 1985 **40** 9–17.
- The American team:
  Rainer, L., Haseltine, W., Patarca, R., Livak, K. J., Starcich, B., Josephs, S. F., Doran, E. R., Rafalski, J. A.,. Whitehorn, E. A., Baumeister, K., Ivanoff, L., Petteway, J. A., Pearson, M. L., Lautenberger, J. A., Papas, T. S., Ghrayeb, J., Chang, N. T., Gallo, R. C. and Wong–Staal, F. Complete nucleotide sequence of the AIDS virus, HTLV-III. *Nature* 1985 **313** 277–284.

The hepatitis B virus

— Genome sequence of the hepatitis B virus:
Galibert, F., Mandart, E., Fitoussi, F., Tiollais, P. and Charnaty, P. Nucleotide sequence of the hepatitis B virus genome (subtype ayw) cloned in *E.coli. Nature* 1979 **281** 646–650.

— Proteins coded by surface and core antigens:
Tiollais, P., Charnay, P. and Vyas, G. N. Biology of hepatitis B virus, *Science* 1981 **213** 406–411.
Stibbe, W. and Gerlich, W. H. Structural relationships between minor and major proteins of hepatitis B surface antigens. *J.Virol.* 1983 **46** 626–628.

— The cycle of viral multiplication:
Tiollais, P. Pourcel, C. and Dejean, A. The hepatitis B virus. *Nature* 1985 **317** 489–495.

— Detection of the hepatitis B virus in blood or hepatocytes by techniques of genetic engineering (see chapter on genetic engineering).

Comments on

— Viroids:
Riesner, A. and Gross, H. J. Viroïds. *Ann. Rev. Biochem.* 1985 **54** 531–564.

— Prions:

Prusiner, S. B. Novel proteinaceous infections particles cause scrapie. *Science* 1982 **216** 136–144.

ONCOGENES

Discoveries of human proto-oncogenes (Weinberg's and Barbacid's teams):

Tabin, C. J., Bradley, S. M., Bargmann, C. I., Weinberg, R. A., Papageorge, A. G., Scolnick, E. M., Dhar, R., Lowry, D. R. and Chang, E. H. Mechanism of activation of a human oncogene. *Nature* 1982 **300** 143–149.

Reddy, E. P., Reynolds, R. K., Santos, E. and Barbacid, M. A point mutation is responsible for the acquisition of transforming properties by the T24 human bladder carcinoma oncogene. *Nature* 1982 **300** 149–152.

General reviews

Bishop, J. M. Cellular oncogenes and retroviruses. *Ann. Rev. Biochem.* 1983 **52** 301–354.

Enrietto, P. J. and Wyke, J. A. The pathogenesis of oncogenic avian retroviruses. *Adv. Cancer Res.* 1983 **39** 269–314.

Hunter, T. and Cooper, J. A. Protein-tyrosine kinases. *Ann. Rev. Biochem.* 1985 **54** 897–930.

GTPasic properties of different ras p21 provided this protein is normal or activated (by mutation of Val–Gly at 12).

McGrath, J. P., Capon, D. J., Goeddel, D. V. and Levinson, A. D. Comparative biochemical properties of normal and activated human ras p21 protein. *Nature* 1984 **310** 644–649.

Cooperative action of ras *and* myc *oncogenes*

Land, H., Parada, L. F. and Weinberg, R. A. Tumorigenic conversion of primary embryo fibroblasts requires at least two cooperating oncogenes. *Nature* 1983 **304** 596–602.

Direct activation in vivo *of an oncogene with the help of a chemical agent by the team of Barbacid*

Zarbl. H., Sukumar, S., Arthur, A. V., Martin-Zanca, D. and Barbacid, M. Direct mutagenesis of *Ha-ras*-1 oncogenes by *N*-nitroso-*N*-methylurea during initiation of mammary carcinogenesis in rats. *Nature* 1985 **315** 382–385.

Does tyrosine play a key role in the control of growth?

(The tumoral viruses which stimulated uncontrolled growth and the factors which stimulate controlled growth phosphorylise tyrosine).

Kolata, G. Is tyrosine the key to growth control? *Science* 1983 **219** 377–378.

Displacement of an oncogene in DNA

The example of Burkitt's lymphoma.

Taub, R., Kirsch, I., Morton, C., Lenoir, G., Swan, D., Tronick, S., Aaronson, S.

and Leder, P. Translocation of the *c-myc* gene into the immunoglobulin heavy chain locus in human Burkitt lympoma and murine plasmacytoma cells. *Proc. Natl. Acad. Sci. USA*. 1982 **79** 7837–7841.

The protein coded by the 'sis' *oncogene has a sequence very close to that of PDGF* (the growth factor derived from blood platelets).

Doolittle, R. F. Simian sarcoma virus oncogene, *v-sis*, is derived from the gene (or genes) encoding a platelet-derived growth factor. *Science* 1983 **221** 275–276.

Waterfield, M. D., Scarce, G. T., Whittle, N., Stroobant, P., Johnson, A., Wasteson, A., Westermark, B., Heldin, C. H., Huang, J. S. and Deuel, T. F. Platelet-derived growth factor is structurally related to the putative transforming protein p28 sis of simian sarcoma virus. *Nature* 1983 **304** 35–39.

INTERFERONS

Action mechanism of interferons

Den Hartog, J. A. J., Wunands, R. A. and van Boom, J. H. Chemical synthesis of pppA2′p5′A2′p5′A an interferon-induced inhibitor of protein synthesis and some functional analogues. *J. Org. Chem.* 1981 **46** 2242–2251.

Sen, G. C. Mechanism of interferon action: progress toward its understanding. *Prog. Nucleic Acid Res. Mol. Biol.* 1982 **27** 105–156.

Lengyel, P. Biochemistry of interferons and their actions. *Ann. Rev. Biochem.* 1982 **51** 251–282.

GENETIC ENGINEERING

General techniques

Series entitled 'Genetic engineering' begun in 1981, edited by Williamson, R. Academic Press, London.

Lucotte, G. *ABC of genetic engineering*, Inter Editions, Paris 1983.

Some examples of results obtained by genetic engineering techniques (synthesis of various proteins, cloning of genes, etc.)

— Human insulin:
Goeddel, D. V., Kleid, D. G., Bolivar, F., Heyneker, H. L., Yansura, D. G., Crea, R., Hirose, T. Kraszewski, A., Itakura, K. and Riggs, A. D. Expression in *Escherichia Coli* of chemically synthesised genes for human insulin. *Proc. Natl. Acad. Sci. USA* 1979 **76** 106–110.

— Human growth hormone:
Lupker, J. H., Roskam, W. G., Miloux, B., Liauzin, P., Yaniv, M. and Louannau, J. Abundant excretion of human growth hormone by recombinant-plasmid-transformed monkey kidney cells. *Gene* 1983 **24** 281–287.

— Factor VIII, four successive publications in the journal *Nature*.
Gitschier, J., Wood, W. I., Goralka, T. M., Wio, K. L., Chen, E. Y., Eaton, D. H., Vehar, G. A., Capon D. J. and Lawn, R. M. Characterisation of the human factor VIII gene. *Nature* 1984 **312** 326–330.

Wood, W. I., Capon, D. J., Simonsen, C. C., Eaton, D. L., Gitschier, J., Keyt, B., Seeburg, P. H., Smith, D. H., Hollingshead, P., Wion, K., Delwart, E., Tuddenham, E. G. D., Vehar, G. A. and Lawn, R. M. Expression of active human factor VIII from recombinant DNA clones. *Nature* 1984 **312** 330–337.

Vehar, G. A., Keyt, B., Eaton, D., Rodriguez, H., O'Brien, D. P., Rotblat, F., Oppermann, H., Keck, R., Wood, W. I., Harkins, R. N., Tuddenham, E. G., Dr Lawn, R. M. and Capon, D. J. Structure of human factor VIII. *Nature* 1984 **312** 337–342.

Toole, J. J., Knopf, J. L., Wozney, J. M., Sultzman, L. A., Buecker, J. L., Pittman, D. D., Kaufman, R. J., Brown, E., Shoemaker, C., Orr, E. D., Amphlett, G. W., Foster, W. B., Coe, M. L., Knutson, G. J., Fass, D. N. and Hewick, R. M. Molecular cloning of a cDNA encoding human antihaemophiliac factor. *Nature* 1984 **312** 342–347.

— Factor IX

de La Salle, H., Altenburger, W., Elkaim, R., Dott, K., Dieterle, A., Driellen, R., Cazenave, J. P., Tolstoshev, P. and Lecocq, J. P. Active γ-carboxylated human factor IX expressed using recombinant DNA techniques. *Nature* 1985 **316** 268–270.

Techniques of genetic engineering in clinical biology

— Principles of techniques used:

Goossens, M. and Kan, Y. Y. DNA analysis in the diagnosis of hemoglobin disorders. *Methods Enzymol.* 1981 **76** 805–817.

— Some examples of applications:

Detection of the DNA of the hepatitis B virus

Brechot, C., Scotto, J. Charney, P., Hadchouel, M., Degos, F., Trepo, C. and Tiollais, P. Detection of hepatitis B virus DNA in liver and serum: a direct appraisal of the chronc carrier state. *Lancet* 1981 **ii** 765–768.

Prenatal diagnosis of sickle-cell anaemia:

Chang, J. C. and Kan, Y. W. A sensitive new prenatal test for sickle cell anaemia. *N. Engl. J. Med.* 1982 **307** 30–32.

Goosens, M., Dumez, Y., Kaplan, L., Lupker, M., Chabret, C., Henrion, R. and Rosa, J. Prenatal diagnosis of sickle cell anemia in the first trimester of pregnancy. *N. Engl. J. Med.* 1983 **309** 831–833.

Microinjections of genes into embryos — e.g. of transgenic mice

Palmiter, R. D., Brinster, R. L., Hammer, R. E., Trumbauer, M. E., Rosenfeld, M. G., Birnberg, N. C. and Evans, R. M. Dramatic growth of mice that develop from eggs microinjected with metallothionein-growth hormone fusion genes. *Nature* 1982 **300** 611–615.

BOOKS

Three excellent books on biochemistry devoting a large part of their text to the biochemistry of nucleic acids and to the expression of the genome are:

Alberts, B., Bray, D., Lewis, J., Raff, M., Roberts, K. and Watson, J. D. *Molecular biology of the cell* 1983, 1221 p.
Lehninger, A. *Principles of biochemistry*, Worth, New York, 1982.
Stryer, L. *Biochemistry*, W. H. Freeman, San Francisco, 949 pp.

REVIEWS

Articles on this subject, usually written by very well-known specialists, can also be found in the following scientific journals:

— *Scientific American*,
— *La Recherche*,
— *Medicine et Sciences,*
— *Nature,*
— *Human Genetics,*
— *Clinical Genetics,*
— *American Journal of Human Genetics,*
— *Science.*

AND TO DIVERT YOU

- Watson, J. D. *The double helix*, Weidenfeld & Nicolson, London, 1981.
- Lwoff, A. and Ulmann, A. (eds), *Les origines de la biologie moleculaire, un hommage a J. Monod*, Etudes vivantes, Paris 1980.
- Danchin, A. *L'oeuf et la poule.* Fayard, 1983.
- Thomas, L. *The medusa and the snail*, Viking Press, New York, 1979.
- Kageyama, M., Nakamura, K., Oshima, T. and Uchida, T. (eds), *Science and Scientists. Essays by Biochemists, Biologists and Chemists*, Japan Scientific Societies Press. Tokyo, 1981.
- Richter, D. (ed) *Women scientists.* Macmillan Press, London 1982.
- Medawar, P. B. (Nobel Prize for Physiology and Medicine, 1960) *Advice to a young scientist*, Harper & Row, New York, 1979.
- Olson, R. W., *The art of creative thinking.* Harper & Row, New York, 1980.

11

Seventy-five exercises and answers relating to various chapters

The questions below are taken from questions posed since 1979 to students of the Saint-Antoine Faculty of Medicine in their PCEM I examinations. (Some of the original questions have not been reproduced here, others have been very slightly modified.)

1979 examinations

Questions

1. Given the following sequence related to a strand of DNA:

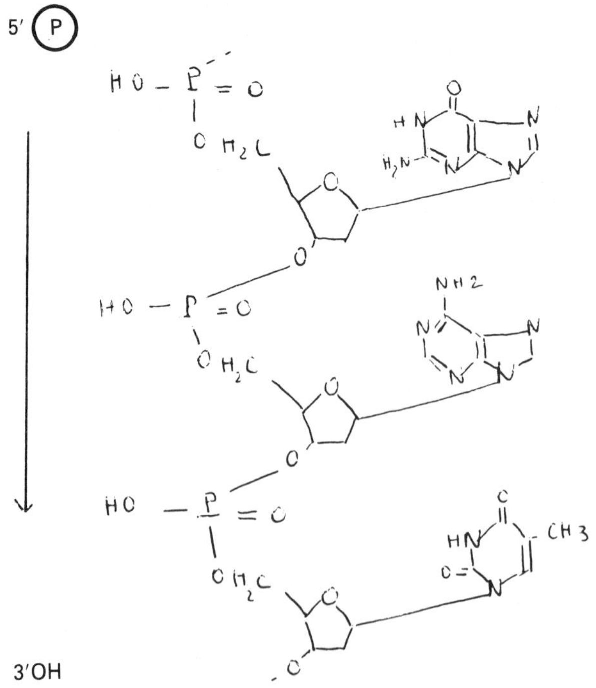

5′ (P)

(Nucleotide x)

(Nucleotide y)

(Nucleotide z)

3′OH

(a) Give the name of the base in nucleotide x.
(b) Give the name of the base in nucleotide y.
(c) Give the name of the base in nucleotide z.
(d) After transcription and translation of this strand of DNA what amino acid(s) will be incorporated into the peptide chain during synthesis of these three nucleotides?

2. What is an anticodon?

Answers

1. (a) Guanine. (b) Adenine. (c) Thymine. (d) Ile.
2. An anticodon is a collection of three nucleotides situated on the tRNA. It is the anticodon which, during translation, is recognised by the mRNA codon and is thus responsible for the choice of amino acid incorporated in the peptide chain during synthesis. The bonds between codon and anticodon are hydrogen bonds between complementary and antiparallel bases. Certain anticodons contain I in the 'wobble' position.

1980 examinations

Questions

1. Explain by what mechanism Endoxan exercises an anticancer action.
2. Take the following mRNA sequence:

In the course of translation the peptide beginning with the following amino acids is obtained:

Met—Ser—Asp—...........

If, by a mutation, the C indicated by the arrow is transformed into U, what peptide sequence will then be obtained?

3. What peptide will be obtained during translation of the synthetic mRNA

poly-AC?

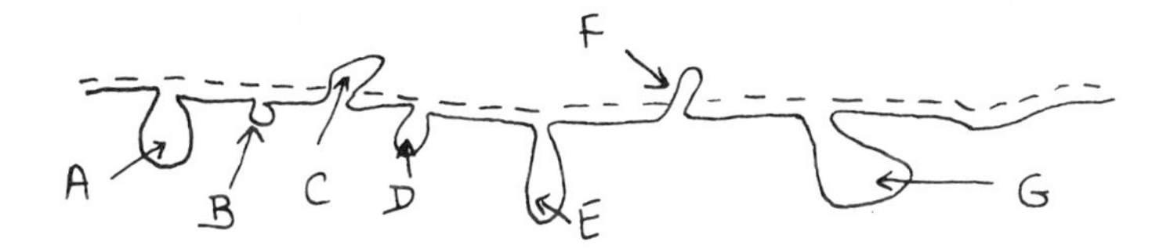

4. This diagram represents what is observed by electron microscopy after a hybridisation experiment (continuous line, a sequence of a strand of ovalbumin DNA; broken line, the corresponding mRNA).
   (a) What do the loops A,B,C,D,E,F,G represent?
   (b) What is the name of the enzyme used in the laboratory to obtain cDNA (DNA obtained by copying the corresponding mRNA)?
   (c) In genetic engineering techniques, to have such a gene transcribed and translated by *E. coli*, which is used: DNA or cDNA?
   Justify your answer.

Answers

1. Endoxan exercises an anticancer action by impeding the replication of DNA. It is one of the alkylants (bivalent). It affixes preferably to guanines. Thus two guanines situated on opposite strands can be united by the intermediation of *Endoxan*, which prevents separation of the two strands, a separation essential for replication. (In normal DNA, the opposite bases are reunited by weak hydrogen bonds).
2. Met—Leu—Asp.
3. Thr—His—Thr—His....
4. (a) Introns. (b) Reverse transcriptase. (c) In genetic engineering techniques cDNA would be used (because even if *E. coli* can transcribe the gene of ovalbumin (including introns and exons) to give a primary transcript, it does not possess the excision–splicing enzyme needed to give the corresponding nRNA.)

1981 examinations

Questions

1. Given the bases:

   1. Adenine, 2. Guanine, 3. Thymine, 4. Cytosine, 5. Uracil.

   Which of the following can pair up in a strand of RNA when it is folded and which hydrogen bonds are formed between complementary bases?

   (A) 1 and 2
   (B) 2 and 3
   (C) 2 and 4
   (D) 2 and 5
   (E) 3 and 4
   (F) 3 and 5
   (G) 1 and 5

2. How many free alcohol functions (i.e. non-esterified) are there in a molecule of DNA (double stranded)?

   (A) 0
   (B) 1
   (C) 2
   (D) As many as there are atoms of phosphorus

(E) Twice as many as there are atoms of phosphorus
(F) Three times as many as there are atoms of phosphorus
(G) Four times as many as there are atoms of phosphorus
(H) As many as there are molecules of pentose
(I) Twice as many as there are molecules of pentose

3. Of the following nucleotide sequences which is (are) that (those) presenting an inverse symmetry ('palindrome') for six pairs of bases

(A) ....GTAAGC....
....CATTCG....
(B) ....CAAGCT....
....GTTCGA....
(C) ....GTTAAC....
....CAATTG....
(D) ....CGTAAC....
....GCATTG....
(E) ....CACGCT....
....GTGCGA....

4. Consider the following substance:

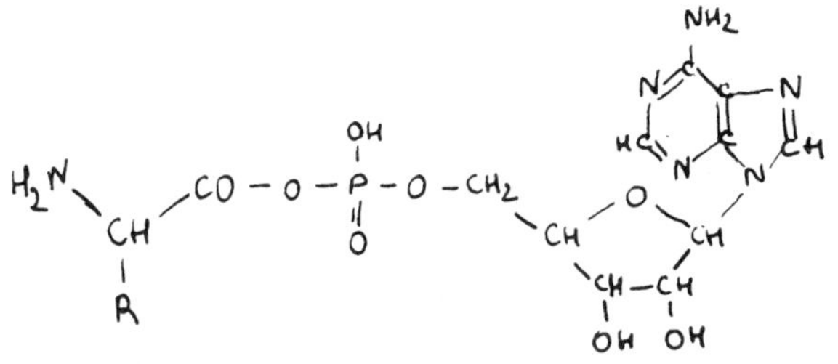

(a) This substance is called:
(A) Cyclic AMP
(B) Dipeptide
(C) Peptide hormone
(D) Amino-acyl AMP
(E) Adenosine monophosphoric acid
(F) Another name

(b) It is present:
(A) In the cytoplasm of eukaryotes
(B) In the chromosomes of eukaryotes
(C) In the ribosomes of eukaryotes

(D) On the cell membrane of eukaryotes
(E) It is never found in eukaryotes

5. Consider the following substance:

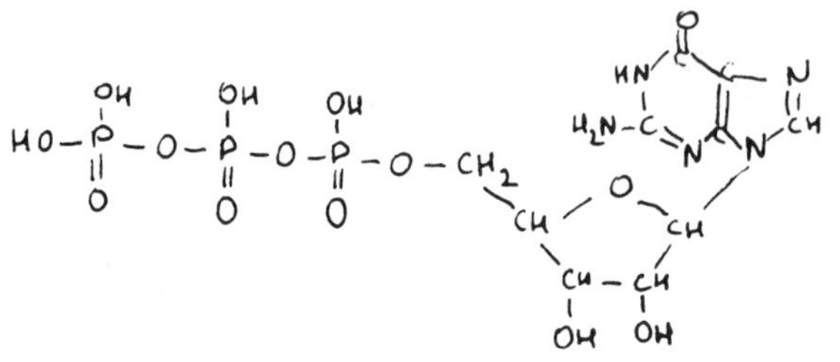

and given:
(1) transcription
(2) replication
(3) translation

(a) Give the name of the substance the formula for which is represented above
(b) Say if this substance is necessary:
(A) In 1 only
(B) In 2 only
(C) In 3 only
(D) In 1 and 2 only
(E) In 1 and 3 only
(F) In 2 and 3 only
(G) In 1, 2 and 3
(H) In none of the three

6. In the lactose operon (*E. coli*), what is the name of the protein substance coded for by the regulator gene in the presence of glucose and in the absence of lactose?

7. A restriction enzyme allows, in well-defined conditions, which of the following to be hydrolysed:
(A) A peptide linkage in a protein in the course of synthesis
(B) A peptide linkage in a protein already synthesised.
(C) An ester linkage between 2 DNA nucleotides.
(D) An ester linkage between 2 mRNA nucleotides.
(E) An ester linkage between 2 tRNA nucleotides.

Answers

1. C, G.
2. C (the OH at the 3′ end of each of the two strands).
3. C.

4. (a) D, (b) A.
5. (a) GTP, (b) G.
6. Repressor.
7. C.

1982 examinations

Questions

1. Consider the following substance:

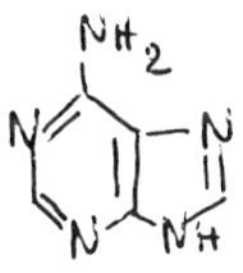

   (a) Give its name.
   (b) With which base can it pair in a molecule of double-stranded DNA?

2. Write the sequence (using capital letters symbolising the nucleotides, and according to the 5′→3′ convention) of the mRNA obtained by transcription of this segment of a strand of DNA.

   DNA sequence: 5′ ATCGTAC 3′   mRNA sequence: 5′ .... 3′

3. What role do enzymes called 'excision–splicing' play?

4. What will be the nucleotide sequence of the anticodon of a tRNA molecule which recognises the three codons: 5′-GGU-3′ 5′-GGC-3′ 5′-GGA-3′ ?

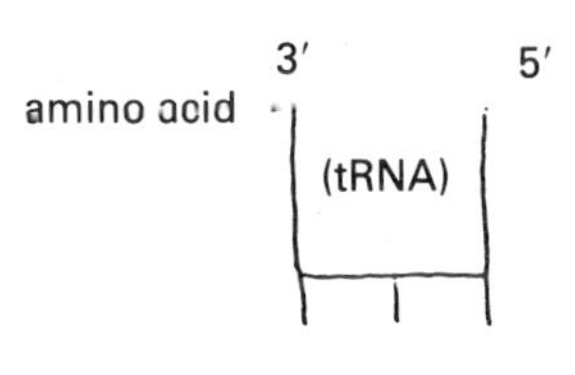

5. There is a very rare disease in man (xeroderma pigmentosa) caused by a genetic deficiency of 'repair enzymes'. Describe the role of these enzymes.

6. Consider the repressor of the lactose operon in *E. coli*.
   (a) Define the chemical nature of this repressor.
   (b) What is the DNA segment to which it is attached called?

7. Consider the following diagram representing preproinsulin:

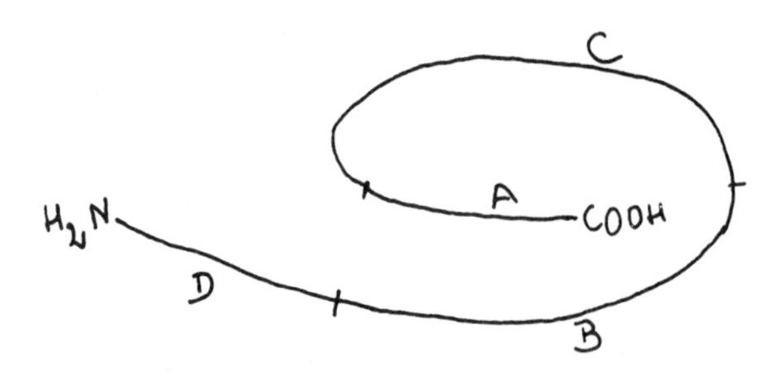

(a) What is the part of the molecule called which is situated at the terminal NH_2 end? Of what chemical units does it largely consist? What is its role?

(b) What post-translational modifications are made to insulin?

8. Why, in genetic engineering techniques, is DNA used which is obtained by occupying the corresponding mRNA (or alternatively artificial genes obtained by synthesis) and not always natural DNA?

Answers

1. (a) Adenine. (b) Thymine.
2. 5′-GUACGAU-3′.
3. They remove the transcriptions of introns contained in a primary transcript and place transcriptions of exons end-to-end.
4. CCI.
5. They facilitate the elimination of nucleotide sequences containing thymine dimers in the DNA.
6. (a) Protein. (b) Operator.
7. (a) Signal sequence. Hydrophobic amino acids, it indicates that the molecule is not destined to remain in the cytosol, but will form part of the membrane proteins or will be exported from the cell. (b) Cutting of the signal sequence, formation of disulphide bridges and elimination of peptide C.
8. Because in the higher eukaryotes the DNA contains introns which will be transcribed and translated by *E. coli*, ending up as a different protein from the protein sought. (*E.coli* does not possess excision–splicing enzymes allowing for the elimination of the transcribed parts which correspond with the introns.)

1983 examinations

Questions

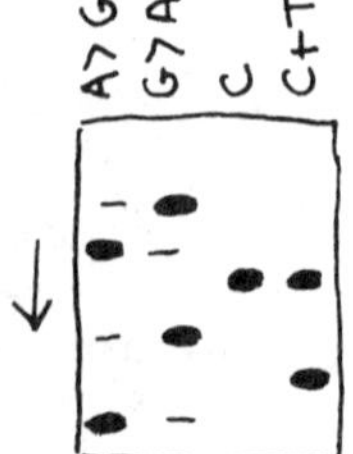

1. The sequence of the segment of the human gene (labelled with ^{32}P at the 5′ end) was determined by the method of Maxam and Gilbert. By examining the attached diagram representing the autoradiogram of a gel where the fragments produced during chemical cleavage have migrated, determine the sequence of bases of the segment of DNA.

   5′……3′

2. Is the left-hand DNA helix (Z-DNA), the mirror image of the right-handed DNA helix. (Briefly justify your answer.)

3. Negram is an antibiotic (used in certain urinary infections) which inhibits the DNA gyrase of the colibacillus. What reaction is promoted preferentially by the DNA gyrase?

4. Suppose (I), a short fragment of cellular proto-oncogene, and (II), the corresponding oncogene, are detected in a human vesical tumour:

   I. 5′...GCCGGCGGT...3′
   3′...CGGCCGCCA...5′ (Strand which will be transcribed)

   II. 5′...GCCGTCGGT...3′
   3′...CGGCAGCCA...5′ (Strand which will be transcribed)

   What will be the sequence of amino acids coded for by:

   (a) The proto-oncogene (I)?

   (b) The corresponding oncogene (II)?

   (In answering this question, assume that the first triplets of each line are in the reading frame.)

5. What is the principle of the synthesis of human insulin by genetic engineering techniques?

Answers

1. 5′-ATGCAG-3′.
2. No. The conformation of the left-handed DNA is different: the helix is more slender, the number of pairs of bases per helical turn is higher, the sugar–phosphate skeleton is zigzag, etc.
3. Catalyses the negative over-twisting of the DNA in the presence of ATP.
4. (a) Ala—Gly—Gly. (b) Ala—Val—Gly.
5. — Chemical synthesis of the two genes corresponding to the A and B chains of human insulin. (The mRNAs of the A and B chains do not exist in the natural state. The natural mRNA is that of preproinsulin.)
   — Incorporation of each gene in a plasmid with transcription signals recognised by *E. coli* above it.
   — Transcription and translation by *E. coli*.
   — Recovery of chains A and B.
   — Assembly of the A and B chains by chemical means (disulphide bridges).

1984 examinations

Questions

1. Consider the base shown:

   (a) What is its name?

   (b) In a molecule of DNA, by what type of bond is it joined to a deoxyribose? (Circle the answer chosen.)
      (A) β-osidic
      (B) Ester
      (C) Anhydrous acid
      (D) Ionic
      (E) Amide
      (F) Hydrogen.

   (c) In the formula above, circle the hydrogen of the base intervening in this linkage with the deoxyribose.

2. Consider a human protein consisting of 600 amino acids.

   (a) What is the minimum number of pairs of nucleotides contained in the structural gene encoding for this protein?

   (b) Why could the number of pairs of nucleotides be greater than that which you have just given? Give three possible reasons.

3. I and II are the two strands of DNA obtained after replication in a human cell. The asterisk represents the methylated bases in strand I.

```
                  *   *
 I = 5'...... A A C G C G A T...... 3'
II = 3'...... T T G C G C T A...... 5' ;
```

   Indicate in strand II, which bases will be methylated by 'maintenance methylases', putting an asterisk just below these bases.

4. Consider the hepatitis B virus. Which part of this virus is responsible for antigenic properties, and is as a result utilised in the preparation of the vaccine against hepatitis B?

5. The following sequence represents the triplets 5, 6 and 7 of exon 1 of the gene coding for the beta globulin, in a normal subject.

(5) (6) (7)

5'.... C C T G A G G A G.... 3' : Strand not transcribed
3'.... (... etc...)T C............ 5' : Strand which will be transcribed

The restriction enzyme MstII which splits the sites 5'-CCTNAGG-3' (N=A, C, G or T), represented by the three arrows x, y, z, on the diagram below:

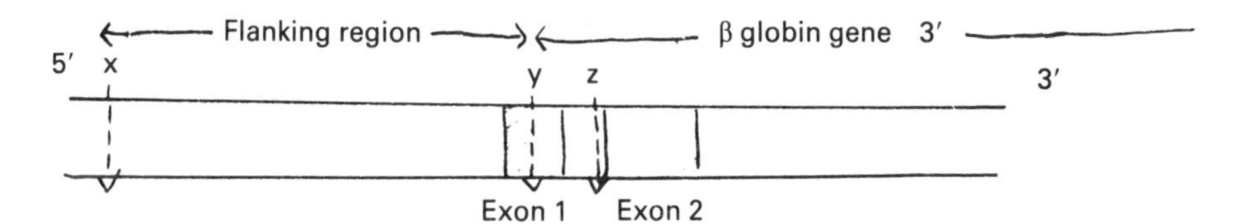

The distance between x and y is 1.15 kb.p and between y and z 0.20 kb.p. (the site y corresponding to triplets 5, 6 and the start of 7 of exon 1).

(a) What will be the sequence of amino acids in the peptide fragment coded by triplets 5, 6 and 7 of the beta globulin gene?

(b) What will be the sequence of amino acids if A of the sixth triplet of the beta globulin gene is mutated to T (the mutation reading to formation of haemoglobin S)?

(c) One of the prenatal diagnostic techniques for sickle-cell anaemia (haemoglobin S) consists of affecting a molecular hybridisation, using for example the restriction enzyme MstII and a radioactive probe which recognises the restriction MstII fragments of the beta globulin gene. Draw the mark(s) which would appear on the autoradiogram obtained at the end of the test in the case of a subject with sickle-cell anaemia (haemoglobin S) in the homozygotic form.

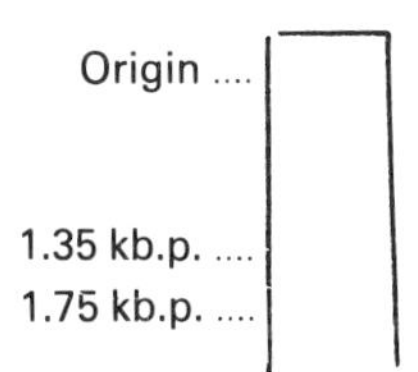

Justify your answer briefly.

Answers

1. (a) Guanine. (b) A. (c) Circle the hydrogen situated on the pentagonal nucleus.
2. (a) $3 \times 600 = 1800$. (b) If there are introns in the structural gene: if the gene codes for a signal sequence (in the case of exported proteins): if the gene codes for a precursor of this protein (the precursor finally being split).
3. Put an asterisk under the two Cs of strand II.
4. The surface protein (HB antigens).
5. (a) Pro–Glu–Glu. (b) Pro–Val–Glu. (c) Draw the mark at the 1.35 kb.p. level. In

sickle-cell anaemia A is replaced by T at site y of the DNA. MstII thus does not cleave at site y. The fragment recognised by the probe will thus have a length of 1.35 kb.p. (in the homozygote form only one mark would be visible).

1985 examinations

Questions

1. Consider the substance whose formula is represented below:

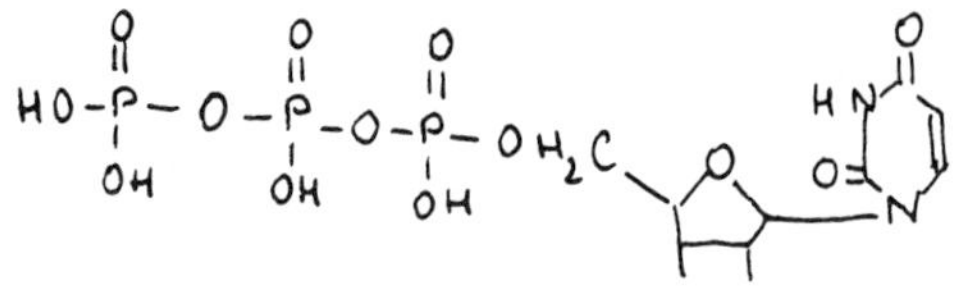

HO-P(=O)(OH)-O-P(=O)(OH)-O-P(=O)(OH)-OH₂C

   (a) What is its name?
   (b) What role does it play?

2. Consider the following sequence of a mRNA (produced by transcription of DNA where a mutation occurred):

5′ 3′
...AUGCCGAACCGGAAACUGACGACCUGACGA

After translation the peptide obtained will begin with Met ..., etc. Show with the help of an arrow, where the translation will now stop (because of the stop codon which has appeared following this mutation). Justify your answer briefly.

3. A new drug is at present being tested to combat the AIDS virus (acquired immunodeficiency syndrome). It would act by inhibiting the 'reverse transcriptase'. Explain briefly what role 'reverse transcriptase' plays in viruses.

4. Consider the tRNA of cysteine:

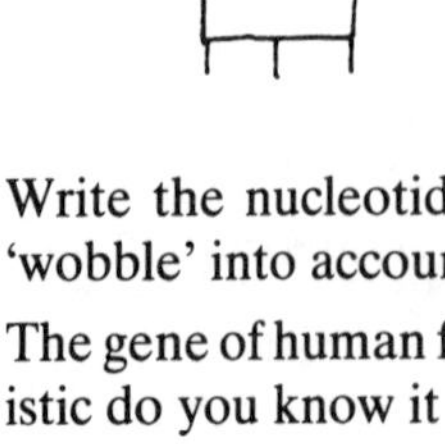

Cys 3′ 5′

Write the nucleotide sequence of its anticodon on the diagram above (taking 'wobble' into account).

5. The gene of human factor VIII has just been isolated; what remarkable characteristic do you know it has?

6. Consider this fragment of gene (of the hepatitis B virus coding for the HB proteins):

```
    5'                       3'
....G A G A A C A T C A C A....     Strand not transcribed
....C T C T T G T A G T G T....     Strand which will be transcribed
    3'                       5'
```

Which peptide sequence will be obtained after transcription and translation? (Give three possible answers depending on the reading frame chosen.)

(a) ...
(b) ...
(c) ...

7. There are two chains of human insulin, A and B.
   (a) In human DNA are there genes coding for each of these two chains? (Comment briefly in your answer.)
   (b) What is the practical significance of this question if one wishes to produce human insulin using techniques of genetic engineering?

Answers

1. (a) UTP. (b) It supplies the UMP and the energy needed for transcription.
2. The arrow should be entered between the sixth and seventh nucleotides from the 3′ end, i.e. just before the stop codon UGA.
   The first stop codon met with should be in phase with AUG. (Take care not to choose stop codon UGA situated 14 nucleotides from end 3′. It is not in phase with AUG.)
3. Reverse transcriptase is an enzyme found in retroviruses which allows these RNA viruses to be copied (by inverse transcription) on a double-stranded DNA. The double-stranded DNA can then be integrated into the DNA of the host cell, thus enabling replication of the viruses to take place.
4. 3′-ACG-5′.
5. This is the longest gene known at present (very many exons).
6. (a) Glu–Asn–Thr. (b) Arg–Thr–Ser. (c) Glu–His–His.
7. (a) No. The DNA contains the preproinsulin gene. (b) Genes corresponding to the A and B chains have to be prepared by chemical synthesis. If the gene of proinsulin were taken, *E. coli* would not be capable of cutting the peptide C of the proinsulin.

1986 examinations

Questions

1. Consider the nucleotide, the formula for which is given below:

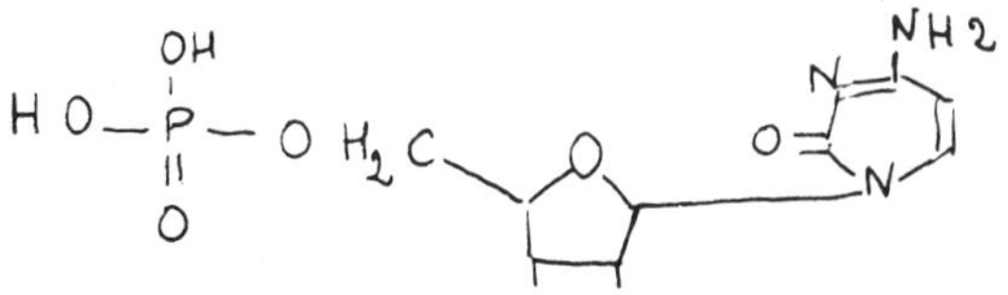

   What is the name of the base which forms part of this nucleotide?

2. Consider the diagram below:

   (a) Indicate on this diagram which base pairs with adenine
      (1) in the case of the DNA molecule, and
      (2) in the case of the RNA molecule.

   (b) By how many hydrogen bonds will this base be linked to adenine:
      (1) in the case of the DNA molecule, and
      (2) in the case of the RNA molecule?

3. Consider a bacterium (*Haemophilus influenzae*) which synthesises the restriction enzyme Hind III. This restriction enzyme recognises the sequence

   AAGCTT
   TTCGAA

   Can such a sequence exist in the DNA of the bacterium without being cut? Justify your answer briefly.

4. Consider the formula shown below:

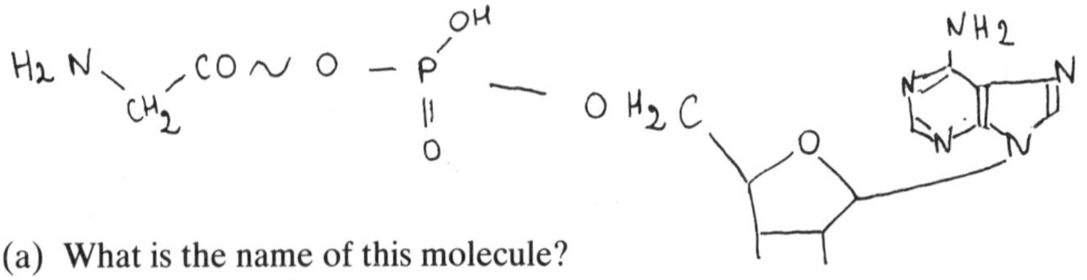

   (a) What is the name of this molecule?

   (b) Cite one known role of this substance.

5. Write (using capital letters to symbolise the nucleotides and the 5′→3′ convention) the mRNA sequence obtained by transcription of this segment of strand of DNA: 5′ CTTAGCGTA 3′.

   Sequence of mRNA: 5′ .... 3′

6. On the diagram below write (using capital letters to symbolise the nucleotides) the sequence of the DNA segment coding for tryptophan:

   5′ ... 3′ Strand of DNA which will not be transcribed
   3′ ... 5′ Strand of DNA which will be transcribed

7. Among the following characteristics pick those which apply to histone proteins:
   - ☐ they are rich in basic amino acids,
   - ☐ they are rich in acid amino acids,
   - ☐ they can acquire negative charges by phosphorylation of serine residues,
   - ☐ they can lose positive charges by acetylation of lysine residues.

8. Is DNA polymerase capable of starting a chain of DNA? Justify your answer briefly.

9. Consider the following tRNA and mRNA:

   Met ~ ACC 3′ 5′ tRNA

   5′ 3′ mRNA

   Write (using capital letters to symbolise nucleotides and consulting the genetic code (earlier in the book)) the sequences of the anticodon and the corresponding codon represented in the above diagram.

10. In genetic engineering techniques:
    (a) What are the chief characteristics of the probe used in a hybridisation reaction?
    (b) Why can loops (represented in the diagram below) be seen by electron microscopy during a hybridisation experiment between a sequence of single-

stranded DNA and the corresponding mRNA?

(c) What are the principal stages of the constitution of a cDNA bank?

Answers

1. Cytosine.
2. (a) T, U. (b) 2, 2.
3. Yes, because the bacteria possess 'modification enzymes' which methylate a base on this palindrome (at the level of their own DNA). This bacterial DNA sequence will then no longer be recognised by restriction enzymes.
4. (a) glycyl-AMP.
   (b) This is an intermediary (activated glycine) in the formation of the tRNA glycyl necessary for the synthesis of proteins (stage of translation).
5. 5′-UACGCUAAG-3′.
6. 5′-TGG-3′
   3′-ACC-5′.
7. Tick squares 1, 3 and 4.
8. No, it cannot initiate a nucleic acid chain. A primer (of RNA) must be synthesised by the RNA polymerase.
9. 3′-UAC-5′ (on the tRNA)
   5′-AUG-3′ (on the mRNA)
10. (a) A strand of nucleic acid (DNA or RNA); complementary to the nucleic acid sequence to be identified; must be able to be recognised (radioactive for example). (b) Because the DNA contains introns which are not recognised by the mRNA (in which the transcripts of introns were eliminated by the excision–splicing enzymes). So only the exons of the DNA and the exons of the mRNA are hybridised. (c) A cDNA bank consists of different bacterial clones into which different cDNAs have been integrated. Isolation of the mRNAs; formation of cDNAs (with reverse transcriptase); insertion of a vector (plasmid for example); introduction of the recombinant DNA into *E. coli* and cloning.

Index